信息技术和电气工程学科国际知名教材中译本系列

Smart CMOS Image Sensors and Applications
智能CMOS图像传感器与应用

［日］ Jun Ohta 著

史再峰 徐江涛 姚素英 译

清华大学出版社
北京

Smart CMOS Image Sensors and Applications 1st Edition/by Jun Ohta/ISNB: 0-8493-3681-2

Copyright@ 2008 by CRC Press.
Authorized translation from English language edition published by CRC Press, part of Taylor & Francis Group LLC; All rights reserved; 本书原版由 Taylor & Francis 出版集团旗下, CRC 出版公司出版, 并经其授权翻译出版。版权所有, 侵权必究。

Tsinghua University Press is authorized to publish and distribute exclusively the Chinese (Simplified Characters) language edition. This edition is authorized for sale and distribution in the People's Republic of China exclusively (except Taiwan, Hong Kong SAR and Macao SAR). No part of the publication may be reproduced or distributed by any means, or stored in a database or retrieval system, without the prior written permission of the publisher. 本书中文简体翻译版授权由清华大学出版社独家出版并仅限在中国大陆地区销售。未经出版者书面许可, 不得以任何方式复制或发行本书的任何部分。

Copies of this book sold without a Taylor & Francis sticker on the cover are unauthorized and illegal. 本书封面贴有 Taylor & Francis 公司防伪标签, 无标签者不得销售。

北京市版权局著作权合同登记号　图字: 01-2014-0803

版权所有, 侵权必究。举报: 010-62782989, beiqinquan@tup.tsinghua.edu.cn。

图书在版编目(CIP)数据

智能 CMOS 图像传感器与应用/(日)太田淳著; 史再峰, 徐江涛, 姚素英译. —北京: 清华大学出版社, 2015 (2022.11 重印)
书名原文: Smart CMOS Image Sensors and Applications
(信息技术和电气工程学科国际知名教材中译本系列)
ISBN 978-7-302-39980-3

Ⅰ. ①智… Ⅱ. ①太… ②史… ③徐… ④姚… Ⅲ. ①图象传感器-教材 Ⅳ. ①TP212

中国版本图书馆 CIP 数据核字(2015)第 086346 号

责任编辑: 文　怡
封面设计: 张海玉
责任校对: 焦丽丽
责任印制: 宋　林

出版发行: 清华大学出版社
　　网　　址: http://www.tup.com.cn, http://www.wqbook.com
　　地　　址: 北京清华大学学研大厦 A 座　　　　邮　编: 100084
　　社　总　机: 010-83470000　　　　　　　　　　邮　购: 010-62786544
　　投稿与读者服务: 010-62776969, c-service@tup.tsinghua.edu.cn
　　质量反馈: 010-62772015, zhiliang@tup.tsinghua.edu.cn
　　课件下载: http://www.tup.com.cn, 010-83470236
印　装　者: 北京鑫海金澳胶印有限公司
经　　销: 全国新华书店
开　　本: 185mm×260mm　　印　张: 11　　　　字　数: 267 千字
版　　次: 2015 年 8 月第 1 版　　　　　　　　　　印　次: 2022 年 11 月第 8 次印刷
定　　价: 49.00 元

产品编号: 055519-02

译 者 序

自鸿蒙而始，人类就是通过眼、耳、鼻、口、舌等器官来感知外界客观世界的，其中 80% 以上的信息来自于视觉。新技术的革命使人类社会开始进入信息时代，被称为"电五官"的传感器是获取自然和生产领域中信息的主要途径与手段。固体图像传感器作为人类视觉器官的延伸，是一种高度集成化的半导体光敏元阵列，具有体积小、工作电压低、性能稳定、畸变小等优点。CMOS 型和 CCD 型是常用的两类固体图像传感器。近年来 CMOS 图像传感器技术发展迅速，在消费电子、机器视觉、智能监控等许多应用领域，已取代 CCD 图像传感器而占据了主流市场，并且更多地应用到生物医学、航空航天等重要领域。智能 CMOS 图像传感器是内部集成了智能处理功能电路的 CMOS 图像传感器，相比传统的图像传感器，具有更加优越的性能，而且可以实现许多传统图像传感器无法实现的功能，应用前景相当广阔。

本书是一本关于智能 CMOS 传感器的特性和应用方面的书。前半部分主要介绍了 CMOS 图像传感器的原理和结构，后半部分主要介绍了智能 CMOS 图像传感器的关键要素及应用。全书通过递进的结构，循序渐进、力求清晰完整地给出智能 CMOS 图像传感器领域的关键要素，并给出了一些比较新颖的智能 CMOS 图像传感器的应用，如涉及信息通信、生物技术和医学方面的几类应用。本书作者 Jun Ohta 博士是日本著名的图像传感器领域的专家，有着多年的研究经验，本书是其多年科研工作的结晶。在电子科学和通信技术飞速发展的今天，引进一些国外原版教材对国内的科研事业是很有好处的。我们翻译此书的目的是促进智能 CMOS 图像传感器的相关知识在高校中的普及和发展，拓宽读者的视野，为他们将来的研究提供参考。本书可以作为微电子及其相关专业高年级本科生、研究生或相关工程技术人员的参考教材，在学习和科研过程中会很有帮助。

本书的翻译工作主要由天津大学 ASIC 设计中心的史再峰、徐江涛、姚素英三位老师共同完成，张晨、周佳慧、王晶波等多位研究生参与了后期的整理、图表翻译等工作，史再峰老师完成了统稿和审校，在此对整个翻译工作团队致以深深的谢意。百年以前，学贯中西的翻译家严复先生就提出了"信、达、雅"的翻译原则和标准。然而，在翻译实践中要想达到这个标准的确是一件极其困难的事情。尽管参与本书翻译工作的整个团队一遍一遍、字斟句酌地去推敲和修正，力求尽可能地把原版书的精髓呈现出来，但由于译者学识和水平有限，书中难免出现一些错误，敬请广大读者不吝赐教和批评指正。

译 者

2015 年 6 月于天津大学

前　言

智能 CMOS 图像传感器主要强调在图像传感器上实现智能的功能和系统应用。某些智能图像传感器已进行商用,还有一些仅是提出概念而已。智能 CMOS 图像传感器领域发展迅速,同时还在促进发明更多的新型传感器。尽管我在这本书中一直努力收集关于智能 CMOS 图像传感器及其应用领域的相关文献资料,然而这个领域实在太广泛了,以至于部分内容难以尽述。此外,该领域的发展实在太迅速了,以至于在成书的过程中新的热点话题就已经出现。不过,我相信这本书已经充分覆盖了智能 CMOS 图像传感器领域的关键要素,所以它对于该领域的研究生和工程师还是会很有帮助的。

这本书的主要结构如下。首先第 1 章对 MOS 成像和智能 CMOS 图像传感器进行了介绍。第 2 章描述了 CMOS 图像传感器的基本要素和光电器件物理的相关知识,同时介绍了典型的 CMOS 图像传感器(如有源像素传感器(APS))的结构。接下来对智能成像器件描述的几个章节构成了本书的主要部分。第 3 章介绍了智能 CMOS 图像传感器的几个主要特征。第 4 章介绍了利用这些特征进行智能成像的技术,如宽动态范围图像检测、目标跟踪和三维测距等。在最后一章(即第 5 章)介绍了几个关于智能 CMOS 图像传感器的应用实例。

这项工作受到了众多先前出版的涉及 CMOS 图像传感器技术的书籍的启发。特别是:A. Moini 的 *Vision Chips*[1] 一书对视觉芯片的结构进行了全面概述;J. Nakamura 的 *Image Sensors and Signal Processing for Digital Still Cameras*[2] 一书全面介绍了此领域近年来的发展成果;K. Yonemoto 的 *Fundamentals and Applications of CCD/CMOS Image Sensors* 一书以及 O. Yadid-Pecht 与 R. Etinne-Cummings 合著的 *CMOS Imagers: From Phototransduction To Image Processing*[4] 一书,对 CCD 和 CMOS 成像器件进行了详尽深入的概述。其中给我印象最深刻的是 K. Yonemoto 的著作,可惜该书只有日文版本。我殷切希望现在的工作能有助于阐明这一领域的研究,作为对 Yonemoto 的书的有益补充。这本书的成书还受到了在相关领域许多其他日本高级研究人员的著作的影响,包括 Y. Takamura、Y. Kiuchi、T. Ando 和 H. Komobuchi 等诸位专家。J. P. Theuwissen 关于 CCD 的著作对本书的形成也很有帮助。

我衷心感谢所有对这本书提供直接或者间接帮助的人。特别是,我在奈良科技研究院的材料科学研究所光学器件实验室的同事 Takashi Tokuda 教授和 Keiichiro Kagawa 教授,他们对本书的重要贡献体现在第 5 章的主要部分。Masahiro Nunoshita 教授在实验室初期总是给予不断的支持和鼓励;实验室秘书 Kazumi Matsumoto 在许多行政事务上给予大力支持和帮助;实验室在读和已毕业的研究生做出了卓有成效的研究。没有他们的努力这本书是无法问世的,我想对这些人致以诚挚的谢意。

关于人工视网膜主题部分,需要感谢视网膜假体项目的负责人 Yasuo Tano 教授、Takashi Fujikado 教授、Tetsuya Yagi 教授和大阪大学的 Kazuaki Nakauchi 博士。同时感谢 Nidek 有限公司的人工视网膜项目组成员 Motoki Ozawa、Shigeru Nishimura 博士、

Kenzo Shodo、Yasuo Terasawa、Hiroyuki Kanda 博士、Akihiro Uehara 博士和 Naoko Tsunematsu。深深感谢该项目顾问委员会成员东京大学的 Ryoichi Ito 名誉教授和大阪大学的 Yoshiki Ichioka 名誉教授。视网膜假体项目由日本新能源与工业技术发展组织（NEDO）的下一代战略技术专项提供资助，还有来自日本卫生与劳动福利部的劳动与健康基金提供的资助。同时感谢人体内窥图像传感器科研项目组的 Sadao Shiosaka 教授、Hideo Tamura 教授和 David C. Ng 博士。这本书的部分工作内容还得到了 STARC（半导体技术社研究中心）的支持，我在这里同时也感谢研究解调 CMOS 图像传感器的 Kunihiro Watanabe 及其合作者。

我第一次进入 CMOS 图像传感器研究领域是在 1992—1993 年，那时我是科罗拉多大学博尔德分校 Kristina M. Johnson 教授的访问研究员。在那里的经历是很令人振奋的，它让我在回到三菱电气集团后的起步研究中受益颇丰。我要感谢在三菱电气集团的所有同事的帮助和支持，包括 Hiforumi Kimata 教授、Shuichi Tai 博士、Kazumasa Mitsunaga 博士、Yutaka Arima 教授、Masahiro Takahashi 教授、Masaya Oita、Yoshikazu Nitta 博士、Eiichi Funatsu 博士、Kazunari Miyake 博士、Takashi Toyoda 等。

真诚地感谢 Masatoshi Ishikawa 教授、Mitumasa Koyanagi 教授、Jun Tanida 教授、Shoji Kawahito 教授、Richard Hornsey 教授、Pamela Abshire 教授、Yasuo Masaki 和 Yusuke Oike 博士允许我在书中使用他们的数据。我从日本图像信息和电视工程师协会的成员 Masahide Abe 教授、Shoji Kawahito 教授、Takayuki Hamamoto 教授、Kazuaki Sawada 教授、Junichi Akita 教授和日本众多图像传感器组的研究人员，包括 Shigeyuki Ochi 博士、Yasuo Takemura 博士、Takao Kuroda、Nobukazu Teranishi、Kazuya Yonemoto 博士、Hirofumi Sumi 博士、Yusuke Oike 博士和其他人员那学到了许多知识。特别想感谢 Taisuke Soda，是他提供给我出版这本书的机会；还要感谢 Pat Roberson、Taylor 和 Francis/CRC 在完成这本书过程中的耐心。没有他们持续不断的鼓励，我是不可能完成这本书的。

我将向 Ichiro Murakami 致以最深切的感谢，他总是激起我在图像传感器和相关主题的研究热情。最后，我想对我爱妻 Yasumi 致以特别的谢意，感谢她在成书的漫长过程中对我的理解与支持。

Jun Ohta
2007 年 7 月于奈良

目 录

第1章 引言 ·········· 1
1.1 综述 ·········· 1
1.2 CMOS图像传感器的简史 ·········· 1
1.3 智能CMOS图像传感器发展简史 ·········· 4
1.4 本书的内容安排 ·········· 6

第2章 CMOS图像传感器技术基础 ·········· 8
2.1 概述 ·········· 8
2.2 光电探测器的基本原理 ·········· 9
2.2.1 吸收系数 ·········· 9
2.2.2 少数载流子的行为 ·········· 9
2.2.3 灵敏度和量子效率 ·········· 11
2.3 智能CMOS图像传感器中的光电探测器 ·········· 12
2.3.1 PN结光电二极管 ·········· 12
2.3.2 光电门 ·········· 18
2.3.3 光电晶体管 ·········· 18
2.3.4 雪崩光电二极管 ·········· 19
2.3.5 光电导探测器 ·········· 19
2.4 PD的积累模式 ·········· 20
2.4.1 积累模式下的电势变化 ·········· 20
2.4.2 势垒描述 ·········· 21
2.4.3 PD中光生载流子的运动 ·········· 22
2.5 基本像素结构 ·········· 24
2.5.1 无源像素传感器 ·········· 24
2.5.2 3T-APS有源像素传感器 ·········· 26
2.5.3 4T-APS有源像素传感器 ·········· 27
2.6 传感器的外设 ·········· 29
2.6.1 寻址 ·········· 29
2.6.2 读出电路 ·········· 31
2.6.3 模拟-数字转换器 ·········· 32
2.7 传感器的基本特性 ·········· 33
2.7.1 噪声 ·········· 33
2.7.2 动态范围 ·········· 35
2.7.3 速度 ·········· 35
2.8 色彩 ·········· 35

2.9　像素共享 36
2.10　像素结构比较 37
2.11　与CCD的比较 38

第3章　智能功能和材料 40

3.1　简介 40
3.2　像素结构 40
　　3.2.1　电流模式 41
　　3.2.2　对数传感器 42
3.3　模拟域操作 43
　　3.3.1　WTA模式 43
　　3.3.2　映射 44
　　3.3.3　电阻网络 44
3.4　脉冲调制 46
　　3.4.1　脉冲宽度调制 47
　　3.4.2　脉冲频率调制 47
3.5　数字处理 52
3.6　硅以外的材料 53
　　3.6.1　绝缘体上硅 53
　　3.6.2　扩展检测波长 55
3.7　非标准CMOS技术结构 57
　　3.7.1　3D集成 57
　　3.7.2　集成光发射器 57
　　3.7.3　通过非标准结构实现颜色识别 59

第4章　智能成像 62

4.1　简介 62
4.2　低光成像 63
　　4.2.1　低光成像中的主动复位 64
　　4.2.2　PFM在低光成像中的应用 64
　　4.2.3　差分APS 65
　　4.2.4　采用Geiger模式APD的智能CMOS图像传感器 66
4.3　高速度 66
　　4.3.1　全局快门 66
4.4　宽动态范围 67
　　4.4.1　宽动态范围的原理 67
　　4.4.2　双重敏感性 68
　　4.4.3　非线性响应 69
　　4.4.4　多次采样 70
　　4.4.5　饱和探测 72
　　4.4.6　亮度扩散 73

- 4.5 解调 ··· 73
  - 4.5.1 解调的原理 ··· 73
  - 4.5.2 相关法 ·· 74
  - 4.5.3 双电荷存储区法 ·· 75
- 4.6 三维测距 ··· 78
  - 4.6.1 时差测距 ··· 79
  - 4.6.2 三角测距 ··· 82
  - 4.6.3 关键值深度法 ·· 83
- 4.7 目标追踪 ··· 84
  - 4.7.1 用于目标追踪的最大值检测法 ·· 85
  - 4.7.2 用于目标追踪的投影法 ··· 85
  - 4.7.3 基于电阻网络及其他模拟处理的目标追踪 ································· 85
  - 4.7.4 基于数字处理的目标追踪 ··· 85
- 4.8 像素与光学系统的专用装置 ·· 87
  - 4.8.1 非正交排列 ·· 87
  - 4.8.2 专用光学系统 ·· 90

第 5 章 应用 ··· 93
- 5.1 引言 ··· 93
- 5.2 信息和通信应用 ··· 93
  - 5.2.1 光学识别标签 ·· 93
  - 5.2.2 无线光通信 ·· 99
- 5.3 生物技术的应用 ·· 104
  - 5.3.1 多模式功能的智能图像传感器 ·· 105
  - 5.3.2 结合 MEMS 技术电位成像 ··· 107
  - 5.3.3 光学和电化学成像的智能 CMOS 传感器 ·································· 108
  - 5.3.4 荧光探测 ··· 110
- 5.4 医学上的应用 ·· 114
  - 5.4.1 胶囊型内窥镜 ··· 114
  - 5.4.2 视网膜假体 ··· 115

- 附录 A 常量表 ··· 123
- 附录 B 光照度 ··· 124
- 附录 C 人眼和 CMOS 图像传感器 ··· 127
- 附录 D MOS 电容的基本特性 ·· 129
- 附录 E MOSFET 的基本特性 ··· 131
- 附录 F 光学格式和分辨率 ·· 134
- 参考文献 ·· 135

第1章 引　　言

1.1 综述

互补金属氧化物半导体(CMOS)型图像传感器近年来已经成为一个广泛研究的课题，并且取得了长足的发展。如今，它已经能与曾经长期占主流的电荷耦合器件(CCD)型图像传感器共同占据市场了。CMOS图像传感器不仅大量用于便携式数码相机、手机摄像头、手持摄像机和数码单反相机等消费类电子产品中，而且已经广泛用于智能汽车、卫星、安保、机器人视觉等领域。近年来，越来越多的CMOS图像传感器出现在了生物技术和医药领域。在这些应用中，多数需要宽动态范围、高速和高灵敏度等先进性能，而另外一些则需应用实时目标跟踪、三维测距等专属功能。这些性能和功能需求对于传统图像传感器来说是难以实现的，即便是采用一些信号处理设备也不足以实现这些需求。然而智能CMOS图像传感器或者片上集成了智能处理功能的CMOS图像传感器就能实现上述功能。

相比于CCD图像传感器成熟的制造工艺，CMOS图像传感器的制造是基于标准CMOS大规模集成电路(LSI)制造工艺的，该特性使CMOS图像传感器可以在芯片内部集成相关功能电路以实现智能的CMOS图像传感器，从而使它比CCD和传统CMOS图像传感器拥有更优越的性能，并可以实现许多传统图像传感器无法实现的功能。

当前，在智能CMOS图像传感器领域的研究主要有两个方向：其一是提升或改进CMOS图像传感器的一些基础特性，如动态范围、速度、灵敏度等；其二是实现一些新型功能，如三维测距、目标追踪和调制光检测等。在这两个领域内，目前已经有多种新型架构、结构以及新型材料被提出并经过验证。

下列的一些术语也与智能CMOS图像传感器相关：具有计算能力的CMOS图像传感器、集成功能图像传感器、视觉芯片、焦平面图像处理等。这些术语除了视觉芯片以外，其他的实际上都是以图像传感器在成像的基础上加入其他功能名称而命名的。而视觉芯片的命名则源于一个由C. Mead和他的同事提出来并加以完善的仿生视觉系统的设备，这个设备将会在以后的章节中加以介绍。接下来，本书将简要叙述CMOS图像传感器的发展历史。

1.2 CMOS图像传感器的简史

1. MOS成像仪的诞生

MOS型图像传感器的历史起源于固态成像仪替代成像管的时代，如图1.1所示。随着固态图像传感器的诞生，图像传感器的4个重要功能得以实现：分别是光探测、光生电荷的积累、积累电荷向读出电路的转移以及扫描模式。这些功能将在第2章详加介绍。基于X-Y轴寻址型硅结光敏器件的扫描功能在20世纪60年代早期被霍尼韦尔公司的S. R.

Morrison 和 IBM 公司的 J. W. Horton 等人提出,前者称之为"光扫描仪"[9],后者称之为"扫描仪"[10]。另外,P. K. Weimer 等人也提出了使用薄膜晶体管(TFTs)[11]来实现带有扫描电路的固态图像传感器。在这些器件中,用于光探测的光电导薄膜将在本书的 2.3.5 节加以介绍。美国国家航空航天局的 M. A. Schuster 和 G. Strul 使用光电晶体管和开关器件实现了 X-Y 坐标寻址[12],并以此成功地在一个 50×50 像素阵列传感器的基础上得到了完整的图像,这将在本书 2.3.3 节中介绍。

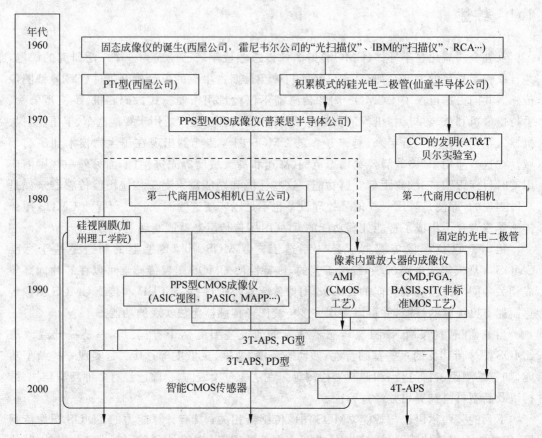

图 1.1 MOS 图像传感器以及相关结构的演变

光生电荷的累积作为 MOS 管图像传感器中重要的功能将在 2.4 节中加以详细介绍。这种结构是由仙童半导体公司的 G. P. Weckler 首先提出的[13]。其中,源极浮空的 MOS 管被用作光电二极管,这种结构至今仍然应用于某些 CMOS 图像传感器中。Weckler 随后使用这种结构制作了一个 100×100 像素阵列的图像传感器加以验证[14]。在这之后,还有许多固态传感器的模型被提出并加以研究[14-17],这些都在参考文献[18]中做了详细总结。

Plessey 公司的 P. J. Noble 提出了改进型固态图像传感器,这种传感器与 MOS 管型图像传感器以及无源像素传感器十分相似,它包括了一个光电二极管和一个具有 X、Y 轴扫描行选的 MOS 管开关以及一个电荷放大器,这些将在本书的 2.5.1 节详细介绍。Noble 还简要阐述了在同一个芯片上集成模式识别的逻辑电路的可能性,这或许就是智能 CMOS 图像传感器的概念首次被提出。

2. 与CCD的对比

IEEE期刊 *Electron Device* 在1968年出版了关于固态图像传感器的论文之后不久，CCD图像传感器诞生了。CCD图像传感器(CCDs)是由AT&T贝尔实验室的W. Boyle和G. E. Smith在1969年发明的，几乎在同时由实验进行了验证。诞生之初，CCD是被当作磁性气泡存储器替代者的半导体气泡存储器而研究的，但不久后它开始应用于图像传感器上。这些早期的CCDs研究成果可以查阅参考文献[21]。

经过科研人员们前赴后继的努力，第一款商业化MOS图像传感器终于在20世纪80年代问世[22-27]。虽然日立公司和松下公司一直在研究MOS图像传感器，但由于CCD图像传感器的画质比MOS型更好，直到最近它还是一直被广泛地制造和使用。

3. 像素内置放大器型固态成像仪

后来，研究人员通过在像素中内置放大器的方法来提高MOS成像仪的信噪比(SNR)。在20世纪60年代，光电晶体管型成像仪被研制成功[12]；80年代后期，科研人员发明了许多放大器型的成像仪，包括电荷调制器件(CMD)[29]、悬浮栅阵列(FGA)型[30]、基于存储型图像传感器(BASIS)[31]、静电感应晶体管(SIT)型[32]、放大器型MOS成像仪(AMI)[33-37]等。除了AMI之外，其他成像仪在制造像素的时候都需要更改标准的MOS制造工艺，因而最终没有被商业化和继续研究。AMI可以使用任意标准CMOS工艺来制造而无须任何改变，而且AMI还使用了和有源像素图像传感器(APS)同样的像素结构。尽管AMI型使用了I-V转换器作为读出电路而APS使用的是源极跟随器，但这并不会造成关键性差异，所以APS也属于像素内置放大器型的图像传感器。

4. 现代CMOS图像传感器技术

有源像素图像传感器最初是由喷气推进实验室(JPL)的E. Fossum等人利用光栅(PG)实现的，后来才使用光电二极管实现[38,39]。使用光栅主要是为了便于处理电荷信号，但是由于作为栅材料的多晶硅在可见光波长区内是不透明的，所以光栅结构的灵敏度较差。现在大多数有源像素图像传感器使用的是光电二极管，这种结构被称为3T-APS(三管有源像素图像传感器)，这种结构已经被广泛地应用在CMOS图像传感器中。在3T-APS的发展之初，其图像质量在固定模式噪声(FPN)和随机噪声方面远不能和CCDs相比。虽然引入了一些噪声消除电路来消除固定模式噪声，但仍无法处理随机噪声的问题。

后来，通过加入一个常用于CCD传感器的具有低暗电流和完全耗尽层的光电二极管，成功地研制出了4T-APS(四管有源像素图像传感器)[40]。4T-APS通过使用相关双采样(CDS)技术成功消除了在随机噪声中占主要成分的 $k_B TC$ 噪声，其成像质量已经可以和CCD图像传感器比肩了。但相比于CCDs，4T-APS最主要的问题还是像素尺寸过大。每个4T-APS像素中有4个晶体管加一个PD和一个浮动扩散区，而每个CCDs中只有一个传输门和一个PD。虽然CMOS制造工艺的进步使像素尺寸在缩小，进一步促进了CMOS图像传感器的发展，但从本质上它还是难以达到CCDs那么小的像素尺寸。最近，科研人员成功研究出一种像素共享技术并已经广泛应用在4T-APS中，从而有效地减小了它的像素尺寸，使得4T-APS已经可以和CCDs相媲美了。图1.2显示了4T-APS像素尺寸的发展趋势,其中CCDs的像素尺寸如图中空心方块所示,该图表明,CMOS图像传感器像素尺寸已经可以与CCD比肩。

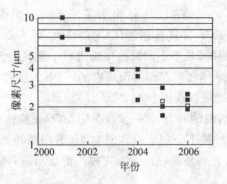

图1.2 4T-APS型CMOS图像传感器像素尺寸发展趋势(图中实心和空心方块分别代表CMOS和CCD图像传感器的像素尺寸)

1.3 智能CMOS图像传感器发展简史

1. 视觉芯片

如图1.3所示,共有3种主要类型的智能图像传感器,它们分别是:像素级智能图像传感器、芯片级(或称为片上相机)智能图像传感器以及列级智能图像传感器。第一种像素级并行处理智能图像传感器也称为视觉芯片。在20世纪80年代,C. Mead和他的同事们在加州理工学院提出并展示了视觉芯片(也称为硅视网膜)[42]。硅视网膜能模仿人的视觉系统,并通过使用大规模硅基集成电路技术来实现大数据量并行处理,该电路工作在亚阈值区

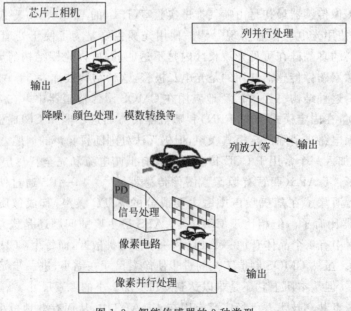

图1.3 智能传感器的3种类型

以实现低功耗的目的,详情见附录 E。此外,通过加入一些二维电阻网络结构的电路,可自动解决某些特定问题,详见 3.3.3 节。而由于光电晶体管(PTr)的增益功能被经常用作光探测器。自从 20 世纪 80 年代以来,Koch、Liu[43]和 A. Moini[1]等人在开发视觉芯片和类似器件上做了大量的工作。目前在焦平面上的大规模并行处理是十分热门的课题,它已经成为了许多相近领域的主要课题,如可编程人工视网膜[44]等。这方面的许多应用已经实现了商业化的目标,如大阪大学的 T. Yagi、S. Kameda 等人应用 3T-APS 研制的双层电阻网络等[45,46]。

图 1.4 是国际半导体技术蓝图(ITRS)中的大规模集成电路的路线图[47]。这幅图向我们展示了动态随机存取存储器(DRAM)半间距的发展趋势,而其他工艺技术如逻辑制程工艺也呈现出几乎相同的发展趋势,这也正好揭示了大规模集成电路的集成密度正如摩尔定律所言:每隔 18~24 个月增长一倍[48]。

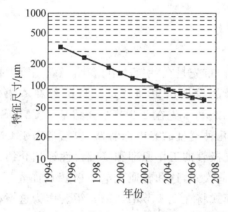

图 1.4 ITRS 发展路线图;DRAM 半间距的发展趋势

CMOS 工艺的进步意味着大规模并行处理或像素级并行处理变得更加切实可行了。这方面已经有相当多的研究成果问世,如基于细胞神经网络(CNN)的视觉芯片[49-52]、可编程多指令流多数据流(MIMD)视觉芯片[53]、仿生学数字视觉芯片[54]、模拟视觉芯片[55,56]等。其他开创性工作包括东京大学和滨松光子学株式会社的 M. Ishikawa 等人研制的基于单指令多数据流(SIMD)处理器的像素级处理视觉芯片等[57-63]。

值得注意的是,一些视觉芯片并不是基于人类视觉处理系统的,因此它们属于像素级处理的范畴。

2. CMOS 工艺技术和智能 CMOS 图像传感器的发展

作为第二种智能传感器,像素级传感器与 CMOS 的工艺技术发展更是息息相关,而与像素并行处理方式几乎无关。这种结构包括片上系统以及相机系统。在 20 世纪 90 年代早期,由于 CMOS 工艺技术的进步,使得实现高集成度 CMOS 图像传感器或用于机器视觉的智能 CMOS 图像传感器变成可能。在这方面前人做了一些开创性的工作,包括 ASIC 视觉系统(最初由爱丁堡大学[64,65]开发后来由 VISI 视觉有限公司接手(VVL))、源于瑞典林雪平大学[67]的接近传感器的图像处理技术(NSIP)(也就是后来的 PASIC[66]),以及由集成视觉产品(IVP)公司研制的 MAPP[68]。

PASIC 可能是第一款使用了列级模数转换器（ADC）的 CMOS 图像传感器[66]。ASIC 视觉系统使用了一种 PPS 结构[67]，而 NSIP 使用的是一个基于传感器的脉冲宽度调制器[67]，MAPP 使用的是 APS 结构[68]，详情见本书的 3.4.1 节。

3. 高性能的智能 CMOS 图像传感器

上文提到的这些图像传感器中包括了一些列并行结构，也就是第三种智能图像传感器结构。因为采用了每列独立供电，列并行结构非常适合于 CMOS 图像传感器。列并行结构可以提高 CMOS 图像传感器的性能，如可以增大动态范围和提高处理速度。通过和 4T-APS 相结合，列并行结构拥有了高质量的成像能力和多种功能。因此，近年来列并行结构被广泛地应用在高性能 CMOS 图像传感器中。大规模集成电路技术的发展也扩大了这种结构的应用范围。

1.4 本书的内容安排

本书的内容安排如下。首先是在"引言"中，介绍了固态图像传感器的总体发展概况，同时介绍了智能图像传感器的发展历史和特点。紧接着在本书的第 2 章中，将详细介绍 CMOS 图像传感器的基础知识，首先在 2.2 节中介绍基于 CMOS 工艺的硅半导体的光电特性。然后在 2.3 节中介绍几种类型的光电探测器，包括常用于 CMOS 图像传感器中的光电二极管等，同时会对光电二极管的工作原理和基础特性详加介绍。在 CMOS 图像传感器中光电二极管工作在累积电荷模式，这和普通光电二极管在如光通信等其他应用中的运行模式有很大不同，这种累积模式将在 2.4 节中加以介绍。作为这章核心内容的像素结构，将在 2.5 节中详细介绍有源像素图像传感器（APS）和无源像素图像传感器（PPS）。在 2.6 节中，将介绍除像素之外的周围电路模块，寻址和读出电路也将在该节中介绍。在 2.7 节中，将讨论 CMOS 图像传感器的基本特性。另外这章中还介绍了色彩（2.8 节）和像素共享技术（2.9 节）。最后我们将在 2.10 节和 2.11 节中作出一些对比分析。

在本书第 3 章中，将介绍一些智能功能和材料，在一些传统 CMOS 图像传感器基础上引入一些新功能的智能 CMOS 图像传感器。首先在 3.2 节中，将介绍一些不同于传统有源像素传感器的像素结构，如对数型传感器。智能 CMOS 图像传感器可以被分为 3 大类：模拟、数字、脉冲，这些内容将在 3.3 节、3.4 节和 3.5 节中加以介绍。CMOS 图像传感器通常基于硅基 CMOS 工艺技术，但其他的一些工艺技术和材料可以被用来实现一些智能功能，例如，蓝宝石硅工艺就是制造智能 CMOS 图像传感器的备选方案。因此我们将在 3.6 节讨论除硅之外的其他工艺材料，而在 3.7 节中将介绍标准 CMOS 工艺以外的其他工艺技术。

结合在第 3 章中介绍的智能功能，第 4 章将会介绍一些智能图像传感器的实例。这些实例中将提到低光成像（4.2 节）、高速度（4.3 节）和宽动态范围（4.4 节）等技术。相比于传统 CMOS 图像传感器，这些智能图像传感器的主要特点就是能提供更好的性能。智能图像传感器的另一个特点是能实现一些传统图像传感器所不具备的多功能性。对于这一点我们也将加以介绍，如解调部分（4.5 节）、三维测距（4.6 节）和目标追踪（4.7 节）。在这章的最后，一些特有的像素和光学排列方式将被介绍。第 4.8 节中将介绍两种智能 CMOS 图像传感器：非正交像素排列型和专用光学型。

在最后的第 5 章中，将介绍一些智能 CMOS 图像传感器的具体应用，涉及信息和通信技术、生物、医药等领域。这些应用都是在近几年问世的，它们无疑会对下一代智能 CMOS 图像传感器产生至关重要的影响。

本书最后的几个附录提供了一些正文中所涉及的部分数据、信息和注释等。

第 2 章　CMOS 图像传感器技术基础

2.1　概述

本章将主要阐述有助于理解 CMOS 图像传感器的相关技术知识。

一个 CMOS 图像传感器通常由一块包含像素阵列的成像区域，行、列寻址电路和读出电路等模块或单元组成，如图 2.1 所示。

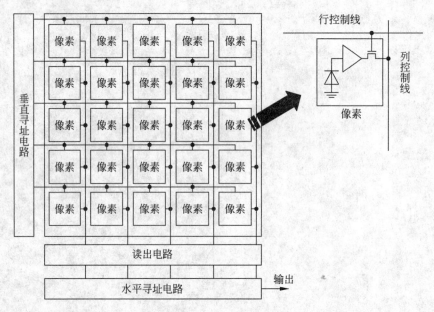

图 2.1　CMOS 图像传感器结构

成像区域是一个二维的像素阵列，每一个像素包含一个光电探测器和多个晶体管。这块区域是图像传感器的核心，成像质量很大程度上取决于该区域的性能。寻址电路用于接通一个像素并读取该像素中的信号值，它一般由扫描器或者移位寄存器来实现，而译码单元则被用来随机访问像素，这种特性有时对于智能传感器来说非常重要。读出电路由一维开关阵列和采样保持（S/H）电路组成，如相关双采样（CDS）这样的降噪电路就在这个区域中。

本章中将详细介绍这些构成 CMOS 图像传感器的基本模块。首先，介绍光电探测器的原理，其中少数载流子的行为起着至关重要的作用。然后介绍几种应用于 CMOS 图像传感器的光电探测器结构，PN 结光电二极管是其中最常用的结构，因此我们将详细介绍它的工作原理和基本特征。此外，也将介绍积累模式，这是 CMOS 图像传感器的一种重要工作模式。其次将介绍基本的像素结构，包括无源像素和有源像素。最后，对 CMOS 图像传感器的其他部分进行介绍，例如扫描单元和译码单元，读出电路和降噪电路等。

2.2 光电探测器的基本原理

2.2.1 吸收系数

如图 2.2 所示,光入射到半导体表面,其中一部分入射光被反射,而其余则被半导体吸收并在半导体内部产生电子-空穴对。这样的电子-空穴对被称作光生载流子。光生载流子的数量取决于半导体材料,并由吸收系数 α 进行描述。

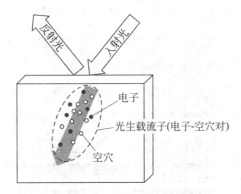

图 2.2 半导体内的光生载流子

α 在这里被定义为:当光入射进半导体的深度为 Δz 时,入射光功率相对减小值 $\Delta P/P$ 的比例,即

$$\alpha(\lambda) = \frac{1}{\Delta z} \frac{\Delta P}{P} \tag{2.1}$$

由式(2.1)可得到下面的关系式:

$$P(z) = P_0 \exp(-\alpha z) \tag{2.2}$$

吸收长度 L_{abs} 定义为

$$L_{abs} = \alpha^{-1} \tag{2.3}$$

需要注意,吸收系数是光子能量 $h\upsilon$ 或波长 λ 的函数,其中,h 和 υ 分别是普朗克常数和光的频率。因此,吸收长度 L_{abs} 的值取决于光的波长。图 2.3 给出了硅材料光吸收系数、吸收长度与入射光波长的对应关系。在 $0.4\sim0.6\mu m$ 波长的可见光范围内,吸收长度位于距离半导体表面 $0.1\sim10\mu m$ 范围内[69]。吸收长度是粗略估算光电二极管结构相关参数的一项重要参数。

2.2.2 少数载流子的行为

入射光在半导体中产生电子-空穴对或光生载流子。P 型半导体中产生的电子是少数载流子,而少子的行为对图像传感器至关重要。举例来说,当入射光为红外线时,由于红外波段光线的吸收长度大于 $10\mu m$(如图 2.3 所示),因此光线可以抵达衬底,所以在 P 型衬底的 CMOS 图像传感器中,衬底中会产生光生少子,即电子。在这种情况下,这些载流子的扩散将大大影响图像传感器的性能,这些载流子可以通过衬底扩散到相邻的光电二极管并导致图像模糊。通常采用红外截止滤光片来消除这一影响,因为红外光可以到达光电二极管较深的区域,即衬底,且相比可见光在衬底产生更多的载流子。

少子的迁移率和寿命凭经验由以下关系给出[70-72],其中包含受主浓度 N_a 和施主浓度 N_d 两个参数:

$$\mu_n = 233 + \frac{1180}{1 + [N_a/(8 \times 10^{16})]^{0.9}} [\text{cm}^2/\text{V} \cdot \text{s}] \tag{2.4}$$

$$\mu_p = 130 + \frac{370}{1 + [N_d/(8 \times 10^{17})]^{1.25}} [\text{cm}^2/\text{V} \cdot \text{s}] \tag{2.5}$$

$$\tau_n^{-1} = 3.45 \times 10^{-12} N_a + 0.95 \times 10^{-31} N_a^2 [\text{s}^{-1}] \tag{2.6}$$

$$\tau_p^{-1} = 7.8 \times 10^{-13} N_d + 1.8 \times 10^{-31} N_d^2 [\text{s}^{-1}] \tag{2.7}$$

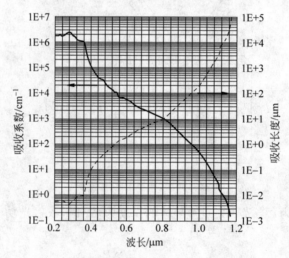

图 2.3 硅材料的吸收系数(实线)、吸收长度(虚线)与波长的关系[69]

根据上述公式,可以用以下关系式估算电子和空穴的扩散长度 $L_{n,p}$

$$L_{n,p} = \sqrt{\frac{k_B T \mu_{n,p} \tau_{n,p}}{e}} \tag{2.8}$$

图 2.4 显示了电子和空穴扩散长度与杂质浓度的关系。当杂质浓度小于 10^{17}cm^{-3} 时,电子和空穴平均可以扩散超过 $100\mu\text{m}$。图 2.5 显示了电子和空穴寿命与杂质浓度的关系。

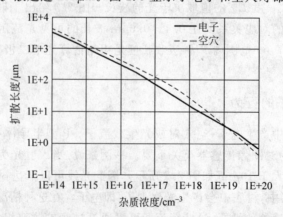

图 2.4 硅材料中的电子和空穴扩散长度与杂质浓度的关系

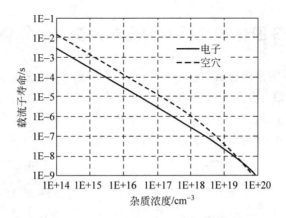

图 2.5 硅材料中的电子和空穴寿命与杂质浓度的关系

2.2.3 灵敏度和量子效率

灵敏度定义为：每单位功率 P_o 的光入射到材料时，所产生的光电流 I_L 的值。它由下式给出

$$R_{ph} \equiv \frac{I_L}{P_o} \tag{2.9}$$

量子效率定义为：光生载流子的数目和入射光子数的比值。每单位时间的入射光子数是 $P_o/(h\nu)$，每单位时间所产生的载流子数目是 I_L/e，因此量子效率可以表示为

$$\eta_Q \equiv \frac{I_L/e}{P_o/(h\nu)} = R_{ph}\frac{h\nu}{e} \tag{2.10}$$

根据式(2.10)，最大灵敏度，即在 $\eta_Q = 1$ 时的灵敏度为

$$R_{ph,max} = \frac{e}{h\nu} = \frac{e}{hc}\lambda = \frac{\lambda[\mu m]}{1.23} \tag{2.11}$$

图 2.6 展示了 $R_{ph,max}$ 随光波长的增加而单调增大，并最终由于材料的禁带宽度 E_g 的限制，在波长为 λ_g 时，其值变为 0。对于硅材料来说，因为其禁带宽度为 1.107eV，所以 λ_g 大约是 1.12μm。

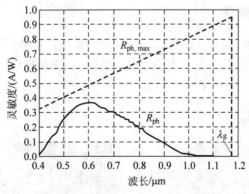

图 2.6 硅材料对应的光灵敏度。实线对应式(2.19)，表示灵敏度 R_{ph} 与波长的关系；虚线对应式(2.11)，表示理想情况下的灵敏度或最大灵敏度 $R_{ph,max}$ 与波长的关系；λ_g 是达到硅禁带宽度限制时所对应的光线波长

2.3 智能 CMOS 图像传感器中的光电探测器

大多数在 CMOS 图像传感器中使用的光电探测器是 PN 结光电二极管(PD),PD 在接下来的部分将详细介绍。在 CMOS 图像传感器中使用的其他光电探测器有光栅(PG)、光电晶体管(PTr)和雪崩光电二极管(APD)。PTr 和 APD 都采用了提高增益的手段,另一种提升增益的探测器是光电导探测器(PCD)。图 2.7 给出了 PD、PG 和 PTr 的结构。

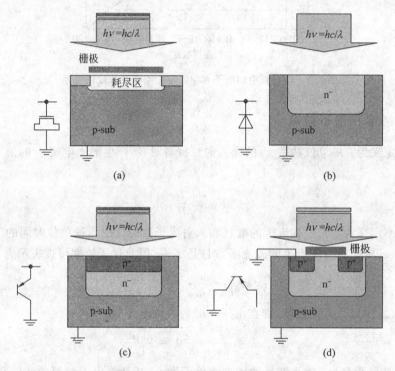

图 2.7 (a)光电二极管符号和结构;(b)光栅的符号和结构;(c)纵向型光电晶体管的符号和结构;(d)横向型光电晶体管的符号和结构

2.3.1 PN 结光电二极管

本节主要介绍一种传统的光电探测器,即 PN 结光电二极管[73,74]。首先解释光电二极管的工作原理,然后讨论如下几个基本特性:量子效率、灵敏度、暗电流、噪声、表面复合和速度。这些特性对智能 CMOS 图像传感器非常重要。

2.3.1.1 工作原理

PN 结光电二极管的工作原理很简单。在 PN 结二极管中,正向电流 I_F 表示为

$$I_{\mathrm{F}} = I_{\mathrm{diff}}\left[\exp\left(\frac{eV}{nk_{\mathrm{B}}T}\right) - 1\right] \tag{2.12}$$

其中 n 为理想因子,I_{diff} 称为饱和电流或扩散电流,它由下式给出

$$I_{\mathrm{diff}} = eA\left(\frac{D_{\mathrm{n}}}{L_{\mathrm{n}}}n_{\mathrm{po}} + \frac{D_{\mathrm{p}}}{L_{\mathrm{p}}}p_{\mathrm{no}}\right) \tag{2.13}$$

其中 $D_{n,p}$，$L_{n,p}$，n_{po} 和 p_{no} 分别是扩散系数、扩散长度、P 型区少数载流子浓度、N 型区少数载流子浓度，A 为 PN 结二极管的横截面面积。PN 结光电二极管中的光电流表示如下：

$$I_L = I_{ph} - I_F$$
$$= I_{ph} - I_{diff}\left[\exp\left(\frac{eV}{nk_B T}\right) - 1\right] \quad (2.14)$$

其中 n 是理想因子。图 2.8 给出了一个 PN 结光电二极管分别在明暗条件下的 I-V 曲线图。如图 2.8 所示，根据不同偏置有三种不同工作模式：太阳能电池模式、光电二极管 (PD) 模式和雪崩模式。

太阳能电池模式：在太阳能电池模式下，光电二极管上不施加任何偏压。在光照下，光电二极管在 PN 结两端产生一个电压，这就像电池一样。图 2.8 显示了开路电压 V_{oc}，开路状态下，开路电压 V_{oc} 可以通过令式(2.14)中 $I_L = 0$ 而得到，所以

$$V_{oc} = \frac{k_B T}{e}\ln\left(\frac{I_{ph}}{I_{diff}} + 1\right) \quad (2.15)$$

这表明，开路电压与入射光强度不呈线性关系。

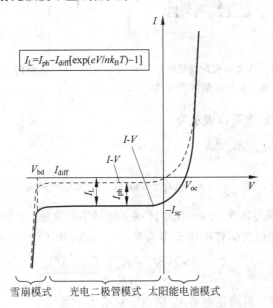

图 2.8 在明暗条件下 PD 的 I-V 特性曲线

光电二极管模式：第二种模式是光电二极管(PD)模式。当二极管反向偏置时，即 $V < 0$ 时，式(2.14)中指数项可以忽略不计，因此 I_L 可近似为

$$I_L \approx I_{ph} + I_{diff} \quad (2.16)$$

这表明，光电二极管的输出电流等于光电流和扩散电流的总和。因此，光电流随光强增加而线性增加。

雪崩模式：第三种模式是雪崩模式。从图 2.8 可以看出，当 PD 强反偏时，光电流急剧增加。这种现象称为雪崩效应，这时电子和空穴发生碰撞电离，从而导致载流子倍增。发生雪崩效应时的电压被称为雪崩击穿电压 V_{bd}，如图 2.8 所示。在 2.3.1.3 节将对雪崩击穿作进一步解释。在 2.3.4 节中将说明雪崩模式会运用于雪崩光电二极管(APD)中。

2.3.1.2 量子效率和灵敏度

利用式(2.2)中吸收系数 $\alpha(\lambda)$ 的定义,可以得到光强表达式:

$$dP(z) = -\alpha(\lambda) P_o \exp[-\alpha(\lambda)z] dz \tag{2.17}$$

为了明确吸收系数取决于波长,将 α 写作 $\alpha(\lambda)$。量子效率被定义为被吸收的光强与总输入光强的比值,因此

$$\eta_Q = \frac{\int_{x_n}^{x_p} \alpha(\lambda) P_o \exp[-\alpha(\lambda)x] dx}{\int_0^\infty \alpha(\lambda) P_o \exp[-\alpha(\lambda)x] dx}$$

$$= \{1 - \exp[-\alpha(\lambda)W]\} \exp[-\alpha(\lambda)x_n] \tag{2.18}$$

式中 W 是耗尽层宽度,x_n 是从材料表面到图 2.9 所示的耗尽区边缘的距离。

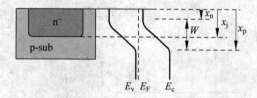

图 2.9 PN 结的结构。该结是在距离表面 x_j 的位置形成的,耗尽区则向 n 型区域 x_n 和 p 型区域 x_p 两侧延伸,因此耗尽区的宽度 W 等于 $x_n - x_p$

利用式(2.18),灵敏度可以表示为

$$R_{ph} = \eta_Q \frac{e\lambda}{hc}$$

$$= \frac{e\lambda}{hc} \{1 - \exp[-\alpha(\lambda)W]\} \exp[-\alpha(\lambda)x_n] \tag{2.19}$$

下面将给出上式中的耗尽区宽度 W、耗尽区在 n 区域中的宽度 x_n 的表达式。利用内建电势 V_{bi} 的概念,并假定所施加的偏置电压大小为 V_{appl},可得到 W 的表达式

$$W = \sqrt{\frac{2\varepsilon_{Si}(N_d + N_a)(V_{bi} + V_{appl})}{eN_a N_d}} \tag{2.20}$$

其中 ε_{Si} 是硅的介电常数,而内建电势 V_{bi} 则由下式给出:

$$V_{bi} = k_B T \ln\left(\frac{N_d N_a}{n_i^2}\right) \tag{2.21}$$

其中 n_i 是硅的本征载流子浓度,$n_i = 1.4 \times 10^{10} \text{cm}^{-3}$。n 型区域和 p 型区域的耗尽层宽度分别为

$$x_n = \frac{N_a}{N_a + N_d} W \tag{2.22}$$

$$x_p = \frac{N_d}{N_a + N_d} W \tag{2.23}$$

图 2.6 给出了硅材料的灵敏度光谱曲线,即灵敏度与光波长的关系。灵敏度光谱曲线和 N 型、P 型区的杂质分布以及 PN 结的位置 x_j 有关。图 2.6 所示的曲线是由如下假设得到的:PN 结的 N 型和 P 型杂质区域是均匀分布的,且为突变结;此外,只统计耗尽区内产

生的光生载流子。实际上耗尽区外产生的一部分光生载流子会扩散进入耗尽区,但是这一部分在此不作统计;而由于长波光线的吸收系数比较低,这种扩散载流子会影响长波灵敏度[75]。另一个假设是忽略了表面复合效应,这部分内容将在2.3.1.4节介绍噪声时再予考虑。这些假设将在2.4节进行讨论。实际图像传感器中的PD表面被SiO_2和Si_3N_4所覆盖,因而量子效率也会受到影响[76]。

2.3.1.3 暗电流

PD中的暗电流有几个来源。

扩散电流:PD本身即存在扩散电流,它可以表示为

$$I_{\text{diff}} = Ae\left(\frac{D_n n_{po}}{L_n} + \frac{D_p n_{no}}{L_p}\right)$$
$$= Ae\left(\frac{D_n}{L_n N_A} + \frac{D_p}{L_p N_D}\right) N_C N_V \exp\left(-\frac{E_g}{k_B T}\right) \quad (2.24)$$

其中A为二极管截面积,N_C和N_V分别是导带和价带的有效状态密度,E_g为禁带宽度。因此,扩散电流随着温度的上升而指数增加。值得注意的是,扩散电流对于偏压的依赖性比较弱,更准确地说,它主要依赖于偏置电压的平方根。

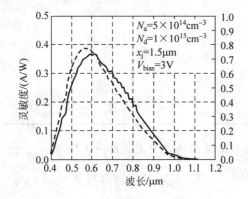

图 2.10 PN结光电二极管的灵敏度(实线)、量子效率(虚线)与波长的关系。PD的参数附于图中

隧穿电流:其他的暗电流还包括隧穿电流、产生-复合电流(G-R电流)、Frankel-Poole电流和表面漏电流[77,78]。隧穿电流包含带带隧穿电流(BTBT)和缺陷辅助隧穿电流(TAT),且与偏置电压呈指数关系,但是对温度依赖性较小[77-79]。如表 2.1 所示,虽然BTBT和TAT引起的暗电流与偏置电压呈指数关系,但是依赖程度不同。当掺杂浓度很大时,隧穿电流变得非常重要,这是因为耗尽区宽度变薄而导致隧穿效应的发生。

产生-复合电流:在耗尽区,载流子浓度降低,并且由于热产生载流子大于复合[77,80],从而产生暗电流。G-R电流由下式给出[77]

$$I_{gr} = AW\frac{en_i}{\tau_g} = AW\frac{e\sqrt{N_C N_V}}{\tau_g}\exp\left(-\frac{E_g}{2k_B T}\right) \quad (2.25)$$

其中,W是耗尽层宽度,τ_g是存在深能级复合中心时的载流子寿命,n_i是本征载流子浓度。这个过程被称为 Shockley-Read-Hall 复合。

碰撞电离电流:随着偏置电压增加,碰撞电离和雪崩击穿会引起暗电流的增加[81,82]。碰撞电离导致的暗电流对于偏置的依赖性是由电子、空穴电离系数α_n和α_p对于电压的依赖

性而引起的,这两个电离系数随着偏压的增加而呈指数式增加。

Frankel-Poole 电流:Frankel-Poole 电流是由于被俘获的电子发射到导带而形成的[77]。和隧穿电流一样,该电流很大程度上依赖于偏置电压。

表面漏电流:表面漏电流由下式给出

$$I_{\text{surf}} = \frac{1}{2} e n_i s_o A_s \tag{2.26}$$

其中 n_i、s_o、A_s 分别是本征载流子浓度、表面复合速率和表面积。

表 2.1 暗电流对于温度和电压的依赖关系[77]。a, a', b, c 是常量

类型	依赖关系
扩散	$\propto \exp\left(-\dfrac{E_g}{k_B T}\right)$
G-R	$\propto \sqrt{V} \exp\left(-\dfrac{E_g}{2k_B T}\right)$
带带隧穿	$\propto V^2 \exp\left(\dfrac{-a}{V}\right)$
缺陷辅助隧穿	$\propto \exp\left(\dfrac{-a'}{V}\right)^2$
碰撞电离	$\alpha \propto \exp\left(\dfrac{-b}{V}\right)$
Frankel-Poole	$\propto V \exp\left(\dfrac{-c}{T}\right)$
表面漏电	$\propto \exp\left(\dfrac{-E_g}{2k_B T}\right)$

暗电流对于温度和偏置电压的依赖关系:对比式(2.24)、式(2.25)和式(2.26),可知不同暗电流对温度的依赖关系不同:只有 I_{surf} 与温度无关,而 $\log I_{\text{diff}}$ 和 $\log I_{gr}$ 分别与 $-\dfrac{1}{T}$、$-\dfrac{1}{2T}$ 呈比例关系。因此,暗电流的来源可以通过不同暗电流成分对温度依赖性的不同来区分。同理,也可通过对偏压依赖性的不同来区分。表 2.1 总结了暗电流对温度和偏置电压的依赖性。

2.3.1.4 噪声

散粒噪声:PD 受散粒噪声和热噪声影响。散粒噪声是由数量为 N 的电子或光子这样的粒子在实际统计上的涨落而引起的。因此 PD 内本身就存在散粒噪声和电子(或空穴)散粒噪声。散粒噪声电流 i_{sh} 的均方根值表示为

$$i_{\text{sh,rms}} = \sqrt{2e \bar{I} \Delta f} \tag{2.27}$$

其中,$\bar{I}$ 和 Δf 分别为信号平均电流和带宽。散粒噪声的信噪比(SNR)可以表示为

$$\text{SNR} = \frac{\bar{I}}{\sqrt{2e \bar{I} \Delta f}} \tag{2.28}$$

因此,当电流或电子数目减少时,SNR 连同散粒噪声一起降低。暗电流同样也会产生散粒噪声。

热噪声:在阻值为 R 的负载电阻中,自由电子存在且随机运动,这种随机运动与负载电阻的温度相关。这种效应导致了热噪声的产生,热噪声也称为约翰森噪声或奈奎斯特噪声。热噪声可以表示为

$$i_{\text{sh,rms}} = \sqrt{\frac{4k_{\text{B}}T\Delta f}{R}} \tag{2.29}$$

在 CMOS 图像传感器中,热噪声以 $k_{\text{B}}TC$ 噪声形式出现,这将在 2.7.1.2 节中进行讨论。

2.3.1.5 表面复合

在传统的 CMOS 图像传感器中,在硅表面与 SiO_2 的交界处存在一些悬挂键,这些悬挂键会产生表面态或界面态,具体表现为非复合中心。一些表面附近产生的光生载流子会被这些中心俘获,它们对光电流没有贡献。因此这些表面态会使得量子效率或灵敏度降低,这叫做表面复合效应。表面复合的特征参数是表面复合速率 S_{surf},表面复合速率和表面处的过剩载流子浓度成正比

$$D_{\text{n}}\frac{\partial n_{\text{p}}}{\partial x} = S_{\text{surf}}[n_{\text{p}}(0) - n_{\text{po}}] \tag{2.30}$$

该复合速率很大程度上依赖于界面态、能带弯曲、缺陷以及其他效应,对于电子和空穴来说大约是 $10 cm^3/s$。对于短波长,如蓝色光,吸收系数较大并且光吸收大部分发生在表面。因此在短波长区域,减小表面复合速率对于实现高量子效率非常重要。

2.3.1.6 速度

近些年来,随着光纤通信和光纤到户(FTTH)技术的发展,硅基 CMOS 光接收器已被广泛地研究和开发。采用 CMOS 技术,包括 BiCMOS 技术的高速光电探测器,在文献[83,84]中进行了详细介绍,应用于 CMOS 光纤通信的高速电路详见文献[85]。

在传统的图像传感器中,光电二极管 PD 的速度并不是一个问题,但是在一些智能图像传感器中需要快速响应的光电二极管。用于无线光纤局域网的智能 CMOS 传感器作为一个例子将在第 5 章介绍,它们基于前面提到过的用于光纤通信的 CMOS 光接收器技术。第 5 章还将介绍另一个例子,一种可以测量飞行时间(TOF)的智能 CMOS 传感器,在这种应用下,APD 被用来实现高速响应。

这里我们考虑 PD 的响应。一般来说 PD 响应的限制因素有 CR 时间常数 τ_{CR}、渡越时间 τ_{tr}、少子扩散时间 τ_{n}(电子)。

CR 时间常数由 PN 结电容 C_{D} 引起,其表达式为

$$\tau_{\text{CR}} = 2\pi C_{\text{D}} R_{\text{L}} \tag{2.31}$$

其中 R_{L} 是负载电阻。

渡越时间的定义为载流子漂移穿过整个耗尽区的时间,它被表示为

$$\tau_{\text{tr}} = W/v_{\text{s}} \tag{2.32}$$

其中 v_{s} 为饱和速率。

产生在耗尽区外的少数载流子可以在经历扩散时间后到达耗尽区,有

$$\tau_{\text{n,p}} = L_{\text{n,p}}^2/D_{\text{n,p}} \tag{2.33}$$

电子的扩散系数为 D_n。

应该注意的是,在渡越时间的限制下,耗尽区宽度 W 和量子效率 η_Q 之间有一个权衡关系。在这种情况下,

$$\eta_Q = [1 - \exp(-\alpha(\lambda)v_s\tau_{tr})]\exp(-\alpha(\lambda)x_n) \tag{2.34}$$

对于 CMOS 图像传感器来说,扩散时间对 PD 响应的影响最大。

2.3.2 光电门

光电门的结构和 MOS 电容相似,当栅极正向偏置时光生载流子积聚在耗尽区。光电门的结构很适合积聚和传输载流子,它已经被应用到一些 CMOS 图像传感器中。光生载流子在 PG 中的积聚如图 2.11 所示,通过施加栅偏压,耗尽区产生并作为一个光生载流子的积聚区域。

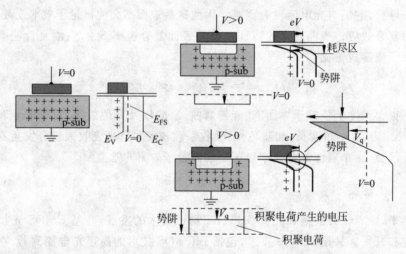

图 2.11 施加栅电压的光电门结构,其产生一个光生载流子积聚的耗尽区

在 PG 中光生区域与表面分开对于一些智能 CMOS 图像传感器是很有用的,这点将在第 5 章中进行讲述。应当指出,PG 灵敏度的缺陷是因为栅极通常由多晶硅制成,而多晶硅是部分透明的并在较短波长或蓝色光区域内具有很低的透射率。

2.3.3 光电晶体管

光电晶体管(PTr)可以由标准 CMOS 技术的寄生晶体管制成。光电晶体管由基极电流增益 β 来对光电流进行放大。因为通过标准 CMOS 工艺技术,基区宽度和载流子浓度不会得到优化,所以 β 值并不高,通常约为 10~20。

特别的是,基区宽度是光电晶体管的权衡因素;当基区宽度增加,量子效率增加但是增益下降[86]。另一个缺点是 PTr 中 β 值变化很大,它会产生固定模式噪声(FPN),详见 2.7.1.1 节。尽管有这些缺点,PTr 还是由于它简单的结构和增益而应用在一些 CMOS 图像传感器中。通过电流镜电路,PTr 可用于电流模式信号处理,这会在 3.2.1 节中讨论。为了解决低光电流下低 β 值的问题,开发了垂直反转层发射源双极型晶体管结构[87]。

2.3.4 雪崩光电二极管

雪崩光电二极管(APD)利用的是光生载流子倍增的雪崩效应。APD具有增益且具备高速响应,因而APD应用于光纤通信和超低光探测,如生物技术。然而,它几乎不用于图像传感器,因为它需要一个超过100V的高压,这样的高压阻碍了APD在标准CMOS技术的使用,除了采用具有CMOS读出电路的APD材料衬底制成的混合图像传感器[88]。增益的变化也在PTr中引起了同样的问题。

由于A. Biber等人的开创性工作,瑞士电子与微电子中心研发了在标准1.2μm BiCMOS技术下的12×24像素的APD阵列[89]。每个像素采用APD控制和读出电路。图像通过制作在19.1V偏压下雪崩倍增为7的传感器获得。

一些关于使用标准CMOS技术制造APD的报告已经发表[90-100],如图2.12所示。在这些报告中,APD偏置在雪崩击穿电压,因此当光子入射到APD时,它迅速打开并产生尖峰电流脉冲。这种现象类似于盖革计数器,因而称其为盖革模式。盖革模式很难在成像中使用,但可以在其他的应用中使用,第5章中对此会有描述。

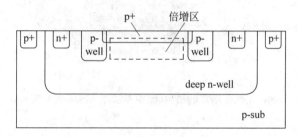

图2.12 采用标准CMOS技术的雪崩光电二极管结构[98]

加州大学圣地亚哥分校的H. Finkelstein等人用0.18μm CMOS技术制造了盖革模式APD[101],他们使用浅沟槽隔离(STI)保护环的APD。他们发现2.5V的偏压足以达到雪崩击穿,这一结果表明,深亚微米技术可以用于制造单个光子雪崩二极管(SPAD)像素阵列的CMOS图像传感器。

2.3.5 光电导探测器

另一种增益检测器是光电导探测器(PCD),它使用光电导效应[86]。一个典型的PCD具有(n^+-n^--n^+)的结构,一个直流偏置施加在两个n^+之间,因此所产生的电场大致局限在n^-区域,这是一个产生电子-空穴对的光电导区,增益来源于空穴的寿命τ_p与电子渡越时间t_{tr}之比,τ_p远大于t_{tr},增益G_{pc}的表达式为

$$G_{pc} = \frac{\tau_p}{t_{tr}}\left(1 + \frac{\mu_p}{\mu_n}\right) \tag{2.35}$$

当光生电子-空穴对通过外加电场被分离,电子会在和空穴复合前多次穿过探测器。值得注意的是,一个较大的增益导致较慢的响应速度,也就是说,在PCD中增益带宽是恒定值,因为增益G_{pc}和载流子寿命τ_p成正比,它将决定探测器的响应速度。最后,PCD通常具有一个相对较大的暗电流,PCD本质上是一个导电装置,会有一些暗电流流过,这可能是作为图像传感器的缺点。根据像素对不同波长(例如X射线、紫外线、红外线)的光响应,一些

PC材料作为探测器覆盖在CMOS读出电路上。雪崩现象发生在部分PC材料上,实现超高纵深比制造技术(HARP),NHK已经开始研发超高灵敏度的显像管[102]。

用于此目的的几种类型的CMOS读出电路已被报道出来,详见文献[103]。PCD的另一个应用是作为片上颜色过滤器的替代物,详见3.7.3节[104-106]。一些PCD也用于快速光电探测器,金属-半导体-金属(MSM)光电探测器也被用于这个目的。

MSM光电探测器是一种PCD,其中一对金属手指被放置在半导体表面上,如图2.13所示[86]。由于MSM结构易于制造,MSM光电探测器也适用于其他材料,如GaAs和GaN。GaAs MSM光电探测器主要用于超高速光电探测器[107],在文献[108,109]中,GaAs MSM光电探测器阵列也用于图像传感器。GaN MSM光电探测器由于其在紫外线区域的灵敏度,已被用于图像传感器的开发[110]。

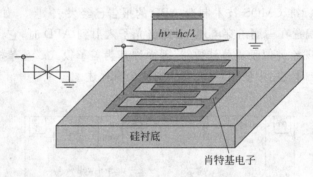

图2.13 MSM光电探测器的结构(图中显示了MSM光电探测器的符号)

2.4 PD的积累模式

CMOS图像传感器中的PD通常工作在积累模式。在这个模式下,PD呈电气浮空状态,当光线照射到PD时,光生载流子生成并由于耗尽区势阱的存在而被扫到表面。PD的积累模式由G. P. Weckler提出并给出证明[13]。电压随着电子的累积而下降,通过测量压降,可以得到光功率的总量。应当指出,电子的积累被解释为由光电流产生的充电电容的放电过程。

我们考虑在一个简单而又典型的情况下,为什么积累模式是CMOS图像传感器所必需的。我们假设以下参数:PD的灵敏度为0.3A/W,面积大小$A=100\mu m^2$,光照强度$=200$lux。假设1lux大概是$1.6\times10^{-7}W/cm^{-2}$,如在附录中描述的,光电流估值为

$$I_{ph} = R_{ph} \times L_0 \times A = 0.3A/W \times 200 \times (1.6\times10^{-7}W/cm^{-2}) \times 100\mu m^2 \approx 10pA$$

虽然它可以测量这种低光电流,但是它很难精确地从一个二维阵列对大量的相同幅度的光电流点以视频的速率测量。

2.4.1 积累模式下的电势变化

PN结光电二极管的结电容C_{PD}为

$$C_{PD}(V) = \frac{\varepsilon_0 \varepsilon_{Si}}{W} \tag{2.36}$$

其中耗尽区宽度 W 取决于外加电压 V,关系式为

$$W = K(V+V_{bi})^{m_j} \tag{2.37}$$

其中 K 是一个常数,V_{bi} 是内建电势差,m_j 是一个有关节形状的参数：突变结 $m_j = 1/2$,线性结 $m_j = 1/3$。C_{PD} 表达式为

$$C_{PD}(V)\frac{dV}{dt} + I_{ph} + I_d = 0 \tag{2.38}$$

其中 I_d 是 PD 的暗电流,利用式(2.36)、式(2.37)和式(2.38),可以得到

$$V(t) = (V_0+V_{bi})\left[1-\frac{(I_{ph}+I_d)(1-m_j)}{C_0(V_0+V_{bi})}\right] \tag{2.39}$$

其中 V_0 和 C_0 分别是 PD 电压和电容的初始值。这个结果显示 PD 的电压几乎是线性下降的,图 2.14 显示了 PD 电压随时间的下降函数,图中 V_{pd} 随着时间增加几乎是线性减少。因此可以根据在一个固定时间内 PD 的压降来估算光强,通常是在 1/30s 的视频帧率。

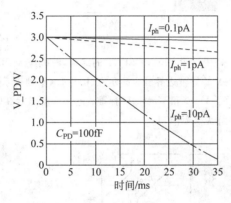

图 2.14 以时间为函数的 PD 的电压降

2.4.2 势垒描述

势垒的描述经常用于 CMOS 图像传感器,因此它是一个重要概念。图 2.15 描述了这个概念。该图中,以 MOSFET 作为实例,源极充当 PD,漏极连接到 V_{dd},源极的杂质浓度比漏极小,栅极处于关闭状态或者处在亚阈值区。

图 2.15(b)显示了沿水平方向的势垒分布,显示导带边缘靠近表面或者表面势垒。此外,图 2.15 中每个区域的电子密度被叠加在势垒分布上,因此很容易看到载流子密度分布如图 2.15(d)所示。载流子密度分布的基线在势垒分布的底部,所以载流子密度往向下的方向增加。值得注意的是,势垒分布或者费米能级可以通过载流子密度来判定；当载流子由入射光生成并积聚在耗尽区中时,势垒深度会根据载流子密度的变化而变化。然而在图像传感器的普通条件下,表面势垒会根据积聚的电荷成比例增加。

图 2.16 显示了一个电气浮动的 PD 势垒描述,这和前面 2.4.1 节是同样的情况。该图中,光生载流子积聚在 PD 的耗尽区,势阱 V_b 由内建电势差 V_{bi} 加上偏压 V_j 得到。图 2.16 显示了当光生载流子收集在势阱中时的积累状态。积累的电荷使势垒深度从 V_b 变成 V_q,如图 2.16(b)所示。V_b 到 V_q 的改变量大概与输入光强和积累时间成正比,如前面 2.4.1 节提到的。

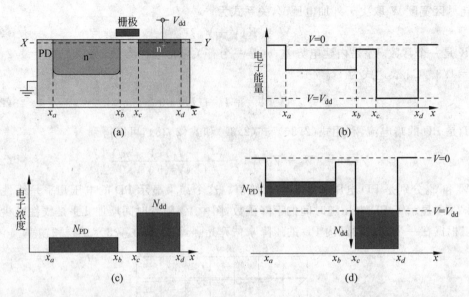

图 2.15 势垒描述的插图。(a)中显示了 n-MOSFET 的结构,源极作为 PD,漏极连接 V_{dd},栅极处于关闭状态;(a)中沿着 X-Y 导带边界分布显示在(b)中,水平轴表示(a)中对应的位置,纵轴表示电子能量,图中显示了 $V=0$ 和 $V=V_{dd}$ 的标准;电子密度显示在(c)中,(d)中是势垒描述,它是(a)与(c)的叠加,详见文献[111]

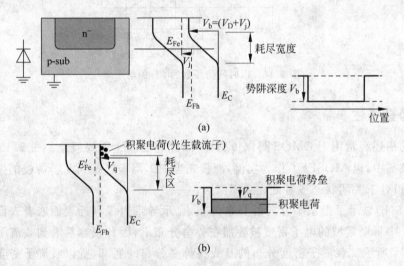

图 2.16 积累前(a)和积累后(b)PD 的势垒描述

2.4.3 PD 中光生载流子的运动

正如 2.3.1.2 节中所解释的,入射光子根据它们的能量或者波长入射到半导体中,较小的光子能量或者较长波长的光子将深入到半导体内,而能量较大或者波长较短的光子则在表面附近被吸收。在耗尽区被吸收的光子直接被电场扫向势阱,并在势阱中积聚,红、绿、蓝三种颜色的光入射到 PD 中,如图 2.17(a)所示,三种光到达不同深度,红光穿透最深,到达 p 衬底区域,它产生少数载流子电子。在 p 衬底区域,电场很小,使得光生载流子只能通过扩散移动，如图 2.17(b)所示。而一些光生载流子在这个区域复合，因而不能增加信号电荷。其他的载流子到达耗尽区的边缘并积聚在势阱中，有助于增加信号电荷。这种贡献程度取决于 p 衬底中产生的电子的扩散长度，扩散长度已经在 2.2.2 节中讨论过了。应当注意，在低杂质浓度区域中载流子的扩散长度较大，因此载流子可以移动很长距离，所以蓝光、绿光还有一部分红光在这种情况下有助于增加信号电荷。

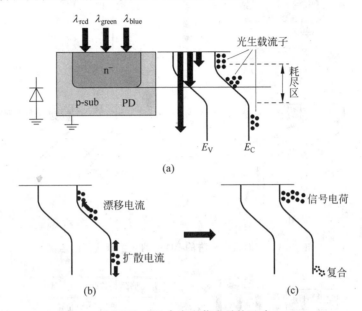

图 2.17　PD 中光生载流子的运动

然而在这种情况下忽略了对载流子有致命伤害的表面态或者界面态。这种状态会在带隙中产生深层能级，在这些状态周围的载流子很轻易地被这些能级俘获。这些状态的寿命一般很长，被俘获的载流子最终会在里面复合，这些被俘获的载流子对信号电荷没有什么贡献。蓝光受到这种效应的影响，比长波的量子效率要小。

扎光电二极管。为了减轻短波长光子量子效率降低的影响，扎光电二极管（PPD）或者掩埋光电二极管（BPD）已经开始发展。从历史上看，PPD 刚开始是应用在 CCD[112,113]上，从 20 世纪 90 年代末开始应用在 CMOS 图像传感器上[40,114-116]。PPD 的结构示意图如图 2.19 所示。PD 顶层表面有一层薄的 p^+ 层，因此 PD 本身看起来是埋在表面下，顶层的 p^+ 薄层的作用是固定表面附近的费米能级，这就是"扎光电二极管"名字的来源。该 p^+ 层和 P 衬底具有相同的电势，因此表面的电势分布强烈弯曲，以便将积累区与具有陷阱状态的表面分离。这种情况下，费米能级是固定的，表面电势也是固定的。

最终，较短波长下的光生载流子被表面附近弯曲的电势分布快速扫向积累区域，有助于信号电荷的增加。PPD 结构有两个优势。首先，PPD 比传统的 PD 具有更小的暗电流，因为表面的 p^+ 层掩盖了造成暗电流主要原因的陷阱。其次，大弯曲的电势分布会产生完全耗尽的累积区，这对 4-Tr 型有源像素传感器是很重要的，这将在 2.5.3 节中讨论。为了实现完全耗尽，不仅需要表面薄的 p^+ 层，也需要一个复杂的精确控制的电势分布设计。近年来，PPD 已被用于高灵敏度的 CMOS 图像传感器。

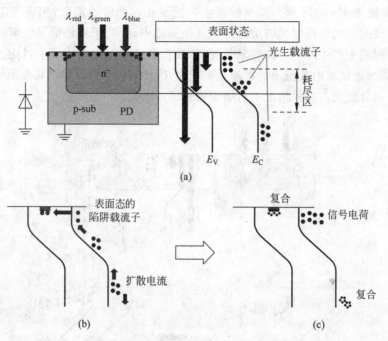

图 2.18 表面具有陷阱的 PD 中的光生载流子的运动

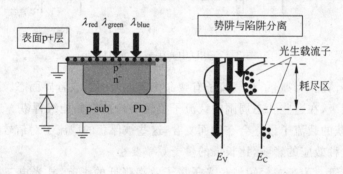

图 2.19 PD 或者扎光电二极管(PPD)表面 p$^+$ 层中光生载流子的运动

2.5 基本像素结构

本节将对基本像素结构作详细说明。从发展历史看,首先开发出来的是无源像素传感器(PPS),然后为了提高图像质量开发出来有源像素传感器(APS)。APS 的一个像素里有 3 个晶体管,而 PPS 只有一个晶体管。为了进一步提高图像质量,现在已经成功研制出来更先进的 APS,它的一个像素内有 4 个晶体管,即所谓的 4T-APS。虽然 4T-APS 大大改善了图像质量,但它的制造工艺很复杂。关于 4T-APS 的实用性目前仍在争议中。

2.5.1 无源像素传感器

无源像素传感器的叫法是为了区别于有源像素传感器,有源像素传感器将在下一小节讨论。首个商用的 MOS 传感器就是一个无源像素传感器[22,24],但是由于信噪比的问题后来终止了研究。PPS 的结构非常简单,一个像素由一个 PD 和一个开关晶体管组成,如图 2.20(a)所示,它类似于动态随机存储器(DRAM)。

由于结构简单,PPS 具有大填充因子(FF),即 PD 截面积与像素面积的比值,一个较大的 FF 更适合于图像传感器。然而,输出信号很容易降低,开关噪声是一个重要问题。在 PPS 发展的第一阶段,积累的信号电荷通过水平输出线转换成读出电流,然后通过电阻或者跨阻放大器转换成电压。但是该方法有如下缺点。

(1) 较大的拖影

拖影一般是以没有任何信号的垂直条纹形式出现的幻影信号。CCD 可以减少拖影。在 PPS 中,拖影发生在信号电荷转移到列信号线时,这个长的水平区间(1H 期间,通常是 64μs)会引起拖影。

(2) 较高的 $k_B TC$ 噪声

$k_B TC$ 噪声是热噪声(详见 2.7.1.2 节),具体来说电荷噪声可表示为 $k_B TC$,其中 C 是采样电容。一个 PPS 在列信号线上具有大采样电容,因此大噪声是不可避免的。

(3) 较高的列 FPN

因为列输出线上的电容 C_C 很大,列开关晶体管需要很大的驱动能力,因此栅极尺寸较大。这会导致大的重叠栅极电容 C_{gd},如图 2.20(a)所示,它将产生大的开关噪声,产生列 FPN。

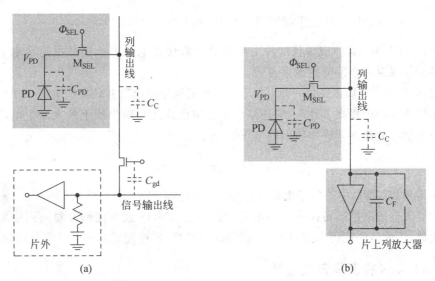

图 2.20 具有两种读出方式的 PPS 的基本像素电路。C_{PD} 是 PD 的 PN 结电容 C_H 与垂直输出线相关联的寄生电容。在电路(a)中,一个片外的功率放大器用于将电荷信号转换成电压信号,而在电路(b)中,列电路中集成了片上电荷放大器,所以信号电荷几乎可以完全读出

为了解决这些问题,横向信号线(TSL)方法被开发出来[117]。图 2.21 显示了 TSL 的概念。在 TSL 结构上,列选择晶体管被用在每个像素中。如图 2.21(b)所示,信号电荷在每一个垂直周期中被选中,它比水平周期短得多,大大减少了拖影。此外,列选择晶体管 M_{CSEL} 需要一个小的采样电容 C_{PD} 而不是标准 PPS 的大电容 C_C,因此 $k_B TC$ 噪声会降低。最终 M_{CSEL} 栅极尺寸得到减小,因此在这种结构中只有很小的开关噪声。为了减小列 FPN,

TSL 结构也被应用在 3T-APS 中。

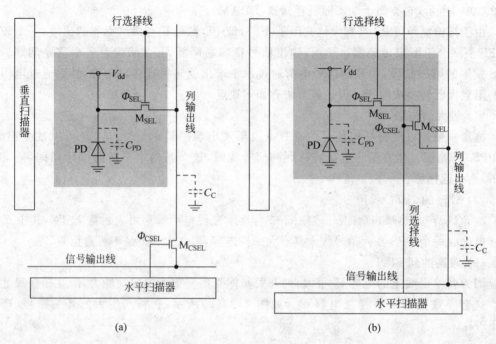

图 2.21 PPS 的改进：(a)传统的 PPS；(b)TSL-PPS[117]

此外，一个带有 MOS 成像仪的片上电荷放大器代替电阻器已经被报道出来了[119]，不过这种结构仅仅对一小部分的像素有效。

目前，电荷放大器放置在每一列上用于完整地提取信号电荷并转换成电压，如图 2.20(b)所示。虽然这种结构提高了性能，它还是由于行输出线或列输出线上的大寄生电容而难以检测到小的信号电荷。列输出线上的电压 V_{out} 为

$$V_{out} = Q_{PD} \frac{C_C}{C_{PD} + C_C} \frac{1}{C_F} \tag{2.40}$$

其中，Q_{PD} 是累积在 PD 中的信号电荷，C_{PD} 是 PD 的电容。

电荷放大器用来精确转换一个小电荷。目前的 CMOS 技术能够在每一列中集成这些电荷放大器，因此信噪比得以提升[120]。值得注意的是，这种配置结构是非常耗电的。

2.5.2 3T-APS 有源像素传感器

APS 得名于它是用有源元件来放大每一个像素的信号，如图 2.22 所示。这种像素结构称为 3T-APS，是与下一节的 4T-APS 相对而言的。一个额外的晶体管 M_{SF} 作为源跟随器，因此输出电压随 PD 电压变化。该信号通过选择晶体管 M_{SEL} 转移到行输出线。与 PPS 相比，APS 提高了图像质量。PPS 是直接将积累信号电荷转移到像素外部，而 APS 是将信号电荷转移到栅极。在这个结构中，电压增益小于 1，电荷增益由积累模式电荷 C_{PD} 和采样保持电荷 C_{SH} 的比例决定。

APS 的工作过程如下。首先，复位晶体管 M_{RS} 开通，接着 PD 复位到 $V_{dd}-V_{th}$，V_{th} 是晶体管 M_{RS} 的阈值电压。接下来，M_{RS} 关闭，PD 进入电气浮动状态。当光入射时，光生载流子

聚集在 PD 结电容 C_{PD} 上。积累的电荷改变了 PD 的势垒，根据输入光强度，PD 电压 V_{PD} 会降低，如 2.4.1 节所述。在一个积累时间之后，比如 33ms 的视频帧率，选通管导通，像素中的输出信号在垂直输出线上被读出。读出过程完成以后，M_{SEL} 关闭，M_{RS} 再次接通，重复上述过程。

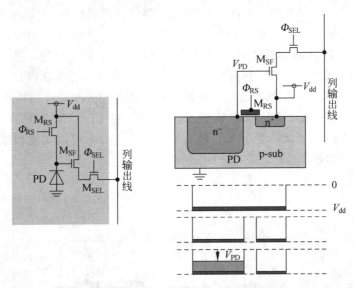

图 2.22　3T-APS 的基本像素电路

值得注意的是，积累的信号电荷并没有被破坏，因此可以多次读取信号。对于智能 CMOS 图像传感器来说，这是非常有用的一个特点。

虽然 APS 克服了 PPS 低信噪比的缺陷，但它依然存在如下一些问题：
(1) 难以抑制的 $k_B TC$ 噪声。
(2) 光探测区域（也就是 PD）同时作为一个光转换区域，这限制了 PD 设计。

这里，我们定义了满阱容量和转换增益。满阱容量是可以累积在 PD 中的电荷数量。满阱容量越大，动态范围(DR)越大。动态范围的定义为最大输出信号值 V_{max} 和可检测的信号值 V_{min} 的比值。

$$DR = 20\log \frac{V_{max}}{V_{min}} [dB] \quad (2.41)$$

转换增益定义为当一个电荷（电子或空穴）积累在 PD 中时的电压变化。因此转换增益等于 $1/C_{PD}$。

满阱容量会随着 PD 结电容 C_{PD} 的增加而增加，而作为根据积累电荷数衡量 PD 电压增长量的转换增益，它和 C_{PD} 成反比。这意味着，满阱容量和转换增益在 3T-APS 中具有相互制约关系。4T-APS 解决了这个制约关系同时抑制了 $k_B TC$ 噪声。

2.5.3　4T-APS 有源像素传感器

为了解决 3T-APS 的问题，研制出来 4T-APS。在一个 4T-APS 中，光探测和光转移区域是分开的，因此累积的光生载流子被转移到浮动扩散区(FD)，在 FD 中载流子将转换成电压。增加一个晶体管用于转移 PD 中积累的电荷到 FD 中，所以单个像素中晶体管总数为 4 个,因而这种像素结构称为四管有源像素传感器(4T-APS)。图 2-23 显示了 4T-APS 的像素结构。

工作过程如下,首先信号电荷积累在 PD 中。假设在初始阶段 PD 中没有积累电荷,满足完全耗尽的条件。在转移积累电荷前通过打开复位晶体管 M_{RS} 使得 FD 复位。复位值被相关双采样读出(CDS)以开启选择晶体管 M_{SEL}。复位读出完成后,积累在 PD 中的信号电荷通过 FD 中的传输门 M_{TG} 转移到 FD 中。重复这一过程,通过打开 M_{SEL} 读出信号电荷和复位电荷。值得注意的是,复位电荷是在信号电荷读出后才被读出的。这个时机对于相关双采样的操作至关重要,可以通过分离电荷存储区(PD)和电荷读出区(FD)来实现;这样可以消除 $k_B TC$ 噪声,而 3T-APS 是无法实现的。通过这种相关双采样的操作,4T-APS 实现了低噪声工作,因此性能可以跟 CCD 媲美。应当指出,在 4T-APS 中,PD 在读出过程必须完全耗尽电荷,因此,这里就需要 PPD。精心设计的电势分布可以实现积累电荷通过传输门到 FD 的完整传输过程。

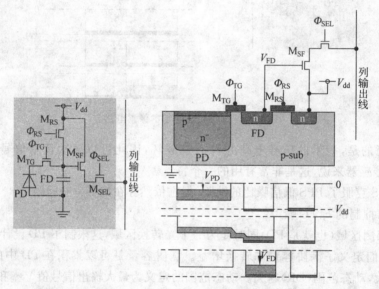

图 2.23 4T-APS 的基本像素结构

4T-APS 的问题

虽然 4T-APS 在低噪声水平上优于 3T-APS,但它还是有以下问题:
(1) 与 3T-APS 相比,增加的晶体管减少了 FF。
(2) 当所积累的信号电荷没有完全转移到 FD 后,可能会出现图像滞后。
(3) 很难对 PPD、传输门、FD、复位晶体管和另外单元的低噪声和低图像滞后性能确立制造工艺参数。

图 2.24 显示了 4T-APS 中不完整的电荷转移。在图 2.24(a)中,电荷完整地转移到 FD 中,而在图 2.24(b)中,部分电荷保留在 PD 中从而引起图像滞后。为了防止不完整的电荷转移,必须精心设计电势分布[121,122]。

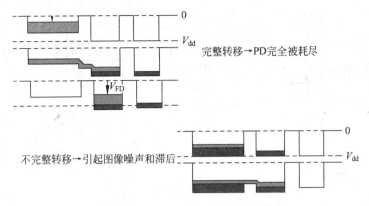

图 2.24 4T-APS 中的不完整电荷转移

2.6 传感器的外设

2.6.1 寻址

在 CMOS 图像传感器中,扫描器或者译码器被用来寻址每个像素。扫描器由一个锁存器阵列或者移位寄存器阵列组成,它根据时钟信号传输数据。当使用扫描器垂直或水平访问时,像素会被顺序寻址。当访问一个任意的像素时,就需要由逻辑门组成的译码器。

译码单元可以使用定制的随机逻辑电路任意地将 N 个输入数据转换成 2^N 的输出。图 2.25 显示了一个典型的扫描器和译码器。图 2.26 给出了一种译码器的例子,它可以解码 3 位输入数据为 6 位输出数据。

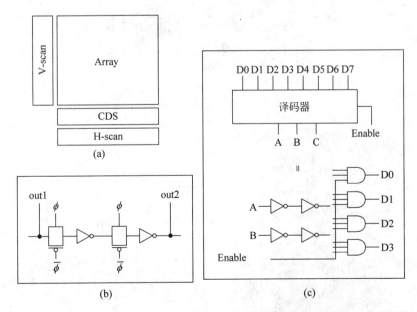

图 2.25 CMOS 图像传感器的寻址方式:(a)传感器结构;(b)扫描器;(c)译码单元

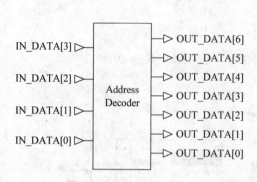

图 2.26　一个译码单元的例子

随机访问。智能 CMOS 图像传感器的一个优点是随机访问能力，其中任意一个像素可以在任何时间被寻址。实现随机访问的典型方法是在每个像素上加一个晶体管，这样像素可以被一个列开关控制，如图 2.27 所示。如上所述，扫描器被行和列地址译码单元替代。应当指出，如果添加额外的晶体管与复位晶体管串联会导致一些时序异常，如图 2.28 所示。在这种情况下，如果 M_{RRS} 打开，在 PD 中积累的电荷分布在 PD 电容 C_{PD} 和寄生电容 C_{diff} 之间，这会降低信号电荷量。

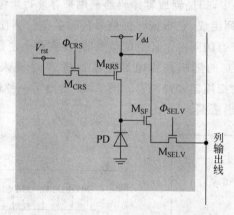

图 2.27　为了随机访问的像素结构

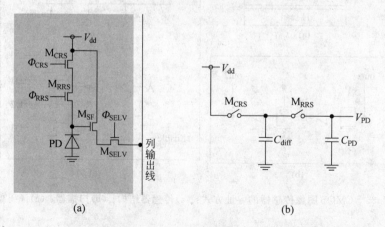

图 2.28　(a)显示在图 2.27 中的不同类型的随机访问的像素结构；(b)等效电路

多分辨率。多分辨率是 CMOS 图像传感器的另一种寻址技术[68,124]。多分辨率是一种使传感器分辨率可变的方法。例如,在 VGA(640×480 像素)的传感器中,可以改变某个因子为原来的 1/4(320×240 像素),改变某个因子为原来的 1/8(160×120 像素),以此类推。当图像后处理负荷低时,为了快速定位传感器中的对象,粗分辨率是一种有效的方法,这对目标跟踪非常有效,如机器人等。

2.6.2 读出电路

2.6.2.1 源极跟随器

PD 的电压用源跟随器读取(SF)。如图 2.29 所示,每个像素中有一个跟随器晶体管 M_{SF},每一列中有一个电流的负载 M_b。在跟随器和负载之间有一个选择晶体管 M_{SEL}。应当指出的是,跟随器的电压增益 A_v 小于 1,表示如下

$$A_v = \frac{1}{1+g_{mb}/g_m} \tag{2.42}$$

其中,g_m 和 g_{mb} 分别是 M_{SF} 的跨导和体效应跨导。源跟随器的 DC 响应与输入不呈线性关系。采样输出电压,并保存在电容 C_{CDS} 中。

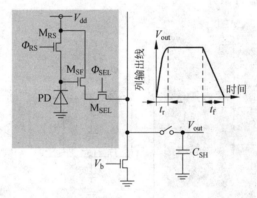

图 2.29 使用源极跟随器的读出电路(插图显示读出周期中输出电压 V_{out} 与时间的依赖关系)

使用源跟随器的读出周期中,S/H 电容 C_{SH} 充电和放电过程是一样的。充电过程中,C_{SH} 在恒定电压模式下充电,因此上升时间 t_r 由恒压模式决定。在放电过程中,C_{SH} 在 SF 恒流源提供的恒流模式下放电,因此下降时间 t_f 由恒流模式决定。图 2.29 对上述过程进行了说明。当读出速度很重要时,就必须考虑这些特性[125]。

2.6.2.2 相关双采样

CDS 用来消除 PD 复位晶体管产生的热噪声,也就是 $k_B TC$ 噪声。在文献[3]中详细描述了几种类型的 CDS 电路。表 2.2 根据文献[3]分类总结了 CDS 的类型。

表 2.2 CMOS 图像传感器的 CDS 类型

类型	方法	特性	参考文献
列 CDS1	一个耦合电容	结构简单但是受列 FPN 影响	[127]
列 CDS2	两个 S/H 电容	结构简单但是受列 FPN 影响	[128]
DDS*	列 CDS 之后	受列 FPN 抑制	[129]
芯片级 CDS	I-V 转换器,芯片上 CDS	受列 FPN 抑制但需要快速工作	[116]
列 ADC	单坡率 ADC	受列 FPN 抑制	[130,131]
	循环 ADC	受列 FPN 抑制	[132]

* 双重三角采样。

图 2.30 给出了一个带有 4T-APS 型像素电路的典型 CDS 电路。基本的 CDS 电路由两组 S/H 电路和一个差分放大器组成。复位电平和信号电平被采样并分别保持在电容 C_R 和 C_S 中,然后对保持在两个电容中的复位值和信号值进行微分得到输出信号。工作原理如图 2.31、图 2.30 的时序所示。在信号读出阶段,选通管 M_{SEL} 从 t_1 到 t_7 一直处于导通状态,Φ_{SEL} 一直处于导通(高电平)。第一步是读取复位电平或者 k_BTC 噪声,然后在 t_2 设置 Φ_{RS} 处于高电平也就是 FD 复位之后,将其存储在电容 C_R 中。采样保持复位信号在电容 C_R 中,Φ_R 在 t_3 变为高电平。接下来读取信号电平。通过在 t_4 时打开传输门 M_{TG} 将积累电荷转移到 FD 中,通过设置 Φ_S 为高电平积累电荷被采样并保持在 C_S 中。最后,通过设置 Φ_Y 为高电平对积累信号和复位信号进行微分。

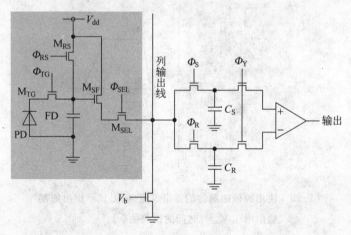

图 2.30 CDS 基本电路

另一种 CDS 电路如图 2.32 所示[127,133]。这种情况下,电容 C_1 用于减去复位信号。

2.6.3 模拟-数字转换器

本节将简单介绍 CMOS 图像传感器中的模拟-数字转换器(ADC)。对于像素少的传感器,如 QVGA(230×320 像素)和 CIF(352×32 像素),一般使用芯片级的 ADC[134,135]。当像素数量增加,会采用列级并行 ADC、逐次逼近 ADC[136,137]、单斜 ADC[66,130,131,138]、循环 ADC[132,139],以及像素级 ADC[61,140,141]。ADC 具体应用的领域与智能 CMOS 图像传感器的结构是一致的,即像素级、列级、芯片级。

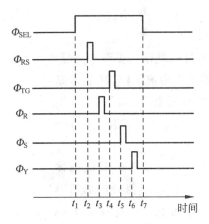

图 2.31 CDS 的时序图（这些符号和图 2.30 中是相同的）

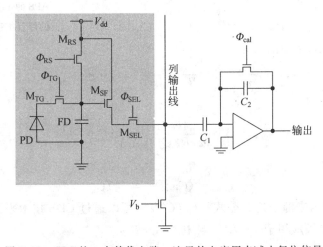

图 2.32 CDS 的一个替代电路。这里的电容用来减去复位信号

2.7 传感器的基本特性

本节将说明一些传感器的基本特性。有关图像传感器的具体测量技术细节请参考文献[142,143]。

2.7.1 噪声

2.7.1.1 固定模式噪声

在图像传感器中，输出信号在空间上的固有变化对图像质量影响很大。这种类型的噪声称为固定模式噪声（FPN）。有规律的变化（如列固定模式噪声）比随机噪声更容易被感知。0.5%的变化是像素固定模式噪声可以接受的阈值，而对于列固定模式噪声来说，0.1%的变化是可以接受的阈值[144]。采用列级放大器有时候会产生列固定模式噪声。在文献[144]中，列输出线和列放大器之间随机化的关系抑制了列固定模式噪声。

2.7.1.2 k_BTC 噪音

在 CMOS 图像传感器中,复位操作主要引起热噪声。当积累电荷通过复位晶体管复位时,热噪声 $4k_BTR_{on}\delta f$ 在积累节点中被采样,其中,δ 是频率带宽,R_{on} 是复位晶体管的导通电阻,如图 2.33 所示。积累节点是 3T-APS 中 PD 的结电容和 4T-APS 中的 FD 电容。

图 2.33 k_BTC 噪声的等效电路。R_{on} 是复位晶体管的导通电阻,C_{PD} 是积累电容,也就是 3T-APS 的 PD 结电容和 4-APS 的浮置扩散电容

热噪声的计算为 k_BT/C_{PD},它不依赖于复位晶体管导通电阻 R_{on},这是因为较大的 R_{on} 阻值会增加每单位带宽热噪声电压,同时减少了带宽[145],这掩盖了热噪声电压与 R_{on} 的关系。现在参考图 2.33 结构得到公式,热噪声电压表示为

$$\overline{v_n^2} = 4k_BTR_{on}\Delta f \tag{2.43}$$

如图 2.33 所示,传输函数为

$$\frac{v_{out}}{v_n}(s) = \frac{1}{R_{on}C_{PD}s+1}, \quad s = j\omega \tag{2.44}$$

因此,噪声公式为

$$\overline{v_{out}^2} = \int_0^\infty \frac{4k_BTR_{on}}{(2\pi R_{on}Cf)^2+1}df = \frac{k_BT}{C} \tag{2.45}$$

电荷的噪声功耗为

$$q_{out}^2 = (Cv_{out})^2 = k_BTC \tag{2.46}$$

"kTC"噪声这个术语来源于这个公式。k_BTC 噪声可以通过 CDS 技术来消除,但它很难被用于 3T-APS,只能被应用于 4T-APS 当中。

2.7.1.3 复位方法

3T-APS 通常的复位操作是通过给 M_{rst} 栅极加高电平或 V_{dd}(如图 2.34(a)所示),并且把 PD 电压 V_{PD} 固定在 $V_{dd}-V_{th}$,其中 V_{th} 是 M_{rst} 的阈值电压。值得注意的是,在复位的最后阶段,V_{PD} 达到 $V_{dd}-V_{th}$,因此 M_{rst} 的栅源电压小于 V_{th},这意味着 M_{rst} 进入亚阈值区域。在这个阶段,V_{PD} 慢慢达到 $V_{dd}-V_{th}$。该复位动作称为软复位[146]。采用 PMOSFET 的复位晶体管可以避免这个问题,因为它有一个 n 阱区域,因此 PMOSFET 相对 NMOSFET 需要更大的面积。相反,在硬复位中,栅极施加的电压大于 V_{dd},因此 M_{rst} 始终高于阈值,从而使复位动作很快完成。在这种情况下,如前所示就会发生 k_BTC 噪声。

软复位会造成图像滞后的缺点,而它的优点是减少 k_BTC 噪声,噪声电压等于 $\sqrt{k_BT/2C}$[146]。通过软复位和硬复位的结合,可以减少噪声并且抑制图像滞后,这被称为刷新复位[147],如图 2.34 所示。在刷新复位时,首先 PD 通过硬复位完全冲刷积累电荷,然后通过软复位减少 k_BTC 噪声。一个刷新复位需要一个开关电路去转换复位晶体管栅极的偏压。

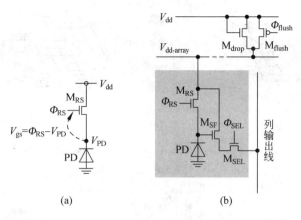

图 2.34 (a)3T-APS 的复位操作和(b)刷新复位[147]

2.7.2 动态范围

图像传感器的动态范围(DR)被定义为输出信号范围与输入信号范围的比值。DR 由本底噪声和满阱容量决定。大部分的传感器几乎具有相同的动态范围(大约为 70dB),主要由 PD 的阱电容量决定。在一些应用中(如汽车),70dB 是不够的,它们需要超过 100dB 的 DR。为了提高 DR,人们已经做出了相当大的努力,详见第 5 章。

2.7.3 速度

APS 的速度根本上是受到扩散载流子的限制。一些在衬底中深区域的光生载流子最终到达耗尽区,作为慢输出信号。电子和空穴的扩散时间与杂质浓度的关系如图 2.5 所示。值得注意的是,电子和空穴的扩散长度为几十微米,有时达到上百微米,为了高速成像需要对其进行小心处理。这种效应大大降低了 PD 的光谱响应,特别是在红外区域。为了减轻这种影响,一些结构被用来防止扩散载流子进入 PD 区域。

CR 时间常数是限制速度的另一个重要因素,因为智能 CMOS 传感器中垂直输出线很长,导致相关电阻和寄生电容较大。

需要指出,与垂直输出线相连的像素中的选择晶体管中的重叠电容总数较大,和垂直输出线的寄生电容相比,它不能被忽略。

2.8 色彩

在传统的 CMOS 图像传感器中识别色彩的方法有三种,如图 2.35 所示。

片上滤色器型。三色的过滤器被直接放置在像素上,通常为红色(R)、绿色(G)和蓝色(B)(RGB)或者为蓝绿色(Cy)、品红色(Mg)、黄色(Ye)和绿色的 CMY 互补色过滤器。CMY 和 RGB 表示(W 代表白色)如下:

$$\begin{cases} Ye = W - B = R + G \\ Mg = W - G = R + B \\ Cy = W - R = G + B \end{cases} \tag{2.47}$$

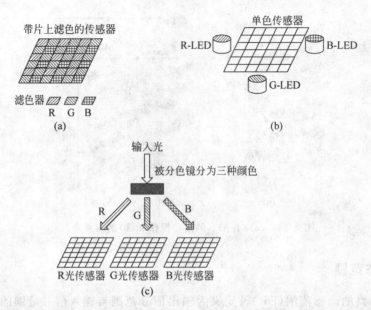

图 2.35 CMOS 图像传感器中识别颜色的方法：(a)片上滤色器型；(b)三成像传感器型；(c)三光源型

Bayer 模式常用来放置 3 个 RGB 过滤器[148]。这种类型的片上过滤器被广泛用于商用 CMOS 图像传感器。通常，该滤色器是有机膜，但有时也用到无机彩色薄膜[149]。控制 α-Si 的厚度以产生颜色反应，这有助于减少滤色器的厚度，对于间距小于 $2\mu m$ 的小间距像素间的光学串扰是很重要的。

三类成像。在三成像方法中，3 个无滤色器的 CMOS 图像传感器被用于 R、G 和 B 颜色。使用两片分色镜将输入光分成三种颜色，这种结构实现了增强色的保真，但需要复杂的光学系统且非常昂贵，它通常用于需要高品质图像的广播系统中。

三类光源。三光源方法使用人工 RGB 光源，每个 RGB 光源对目标进行顺序照明。一个传感器获取 3 种颜色的 3 幅图像，最终的图像需要 3 幅图像的结合。这种方法主要用于医疗内窥镜。颜色保真度是非常好的，但获得整个图像的时间比上述两种方法都要长。这种类型的颜色表示不适用于常规 CMOS 图像传感器，因为它们通常有一个滚动快门，这将在 5.4 节中讨论。

虽然颜色是常规 CMOS 图像传感器中的一个重要特性，智能 CMOS 图像传感器的实现方法几乎是一样的，对其详细的讨论已经超出了本书的范围。一般传感器颜色的处理在文献[2]中。3.7.3 节中介绍了使用智能功能实现彩色的主题。

2.9 像素共享

在像素中的一些部分(例如 FD)可以彼此共享，使得像素尺寸可以减小[150]。图 2.36 显示了像素共享方案的一些例子。在图 2.36 中 FD 驱动共享技术[153]被用于减少 4T-APS 中晶体管数目。通过复位晶体管改变像素漏端电压，从而控制 FD 的电势，以此消除选择晶体管。近年来，使用像素共享技术的 $2\mu m$ 像素间距的传感器已经被报道出来了[155,156]。在文献[156]中，一个 Z 字形 RGB 像素排列改进了像素共享的结构，如图 2.37 所示。

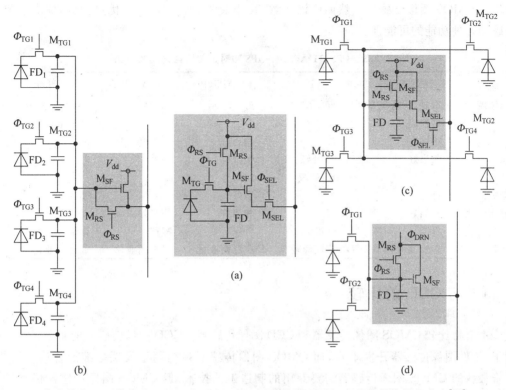

图 2.36　像素共享：(a)传统的 3T-APS；(b)一个选择晶体管和源级跟随器晶体管的共享[151]；(c)当其他部分共享(包括 FD 在内)只有一个 PD 的像素和传输门晶体管[152]；(d)与(c)中一样,但是复位电压受控制[153]

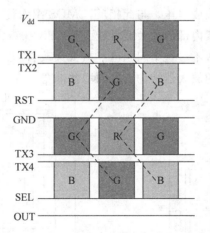

图 2.37　具有 Z 字形的 RGB 像素的像素共享

2.10　像素结构比较

本节中，表 2.3 总结了几种类型的像素架构：PPS、3T-APS 和 4T-APS，以及将在第 3 章中详细讨论的对数传感器。现在，4T-APS 在噪声特性方面具有最佳性能，并将最终广泛应用于 CMOS 图像传感器。然而应该注意，其他系统也具有各自的优点，它们提供了智能传感器各种功能的可能性。

表 2.3 PPS、3T-APS、4T-APS 和对数传感器之间的对比

	PPS	3T-APS	4T-APS(PD)	4T-AP(PG)	Log
敏感度	取决于电流放大器性能	好	好	相当好	好，但是在低光照水平下差
面积消耗	很好	好	相当好	相当好	差
噪声	相当好	相当好（没有KTC减小）	很好	很好	差
暗电流	好	好	很好	好	相当好
成像惰性	相当好	好	相当好	相当好	差
工艺	标准	标准	特殊	特殊	标准
注释	很少商业化	广泛商业化	广泛商业化	很少商业化	最近商业化

2.11 与 CCD 的比较

本节将比较 CMOS 图像传感器和 CCD 图像传感器。CCD 图像传感器的制造工艺仅仅用于 CCD 图像传感器开发本身，而 CMOS 图像传感器最初用于发展标准混合信号工艺。尽管最新的 CMOS 图像传感器需要专用的制造工艺技术，但 CMOS 图像传感器仍然是基于标准混合信号工艺的。

CCD 和 CMOS 图像传感器的结构主要有两个区别，信号传输方法和信号读出方法。图 2.38 显示了 CCD 和 CMOS 图像传感器的结构。CCD 将信号电荷转移到输出信号线的底端并通过一个放大器转换成电压。与此相反，CMOS 图像传感器在每个像素处直接将信号电荷转换成电压。在像素内部的放大器可能引起 FPN 噪声因此早期的 CMOS 图像传感器成像质量比 CCD 差很多，不过这个问题已经被大大改善。在高速操作时，像素内放大器的结构可以获得比芯片级的放大器结构更好的增益带宽。

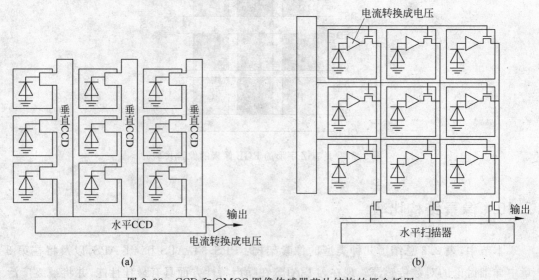

图 2.38 CCD 和 CMOS 图像传感器芯片结构的概念插图

在 CCD 图像传感器中,信号电荷同时传输,这会带来低噪声和高功耗。另外,信号传输在任意时间给每个像素提供了相同的积累时间。与此相反,CMOS 图像传感器中每个像素中的信号电荷会被转换,并且所得信号被逐行读出,因此任意时间在不同行中的像素的积累时间是不同的,这被称为"滚动快门"。图 2.39 显示了滚动快门的起源。三角形状物体从左向右移动,在成像平面上,物体被逐行扫描。在图 2.39(a) 时间点 $Time_k (k=1,2,3,4,5)$,采样点如图是第 1~5 行。原图(图 2.39(b) 左)被扭曲成检测到的图像(图 2.39(b) 右),它是由图 2.39(b) 中对应的点构成。表 2.4 总结了 CCD 与 CMOS 之间的比较,包括这些特征。

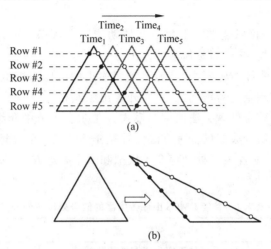

图 2.39 插图显示滚动快门的起源:(a)中三角物体从左向右移动;(b)原始图像被扭曲

表 2.4 CCD 图像传感器和 CMOS 图像传感器的对比

条目	CCD 图像传感器	CMOS 图像传感器
读出方案	一个片上 SF,限制速度	每个列中有 SF,可以显示 FPN 列
同时性	每个像素同时读出	每列顺序复位,滚动快门
晶体管隔离	反向偏置 PN 结	LOCOS/STI*,可以表示压力引起的暗电流
栅氧厚度	厚的以完成完整的电荷转移 ($>50nm$)	薄的由于高速晶体管和低阈值电源电压 ($<10nm$)
栅电极	重叠第一层和第二层多晶硅	多晶硅
隔离层	由于要抑制光波导而薄	厚(约 $1\mu m$)
金属层	通常一层	超过三层

* LOCOS—局部硅氧化;STI—浅沟道隔离。

第 3 章 智能功能和材料

3.1 简介

CMOS 图像传感器可以在芯片上集成智能功能。本章涵盖了几种智能功能及相关材料,首先描述了智能功能,接着讨论像素的输出。

在传统的 CMOS 图像传感器中,从源跟随器输出的电压为模拟信号,如 2.6.2.1 节所述。无论如何,为了实现某些智能功能,人们开发出了其他几种模式,如表 3.1 中总结的使用模拟、数字以及脉冲信号处理方法的电流模式。在 3.2.1 节中介绍了两种类型的电流操作模式,随后讨论了这两种模式的处理方法。首先介绍了模拟的处理方法,然后详细介绍了脉冲模式的处理方法。脉冲模式处理综合了模拟和数字处理方法,最后,介绍了数字处理模式的细节。

表 3.1 智能 CMOS 图像传感器的信号处理方法分类

类 别	优 点	缺 点
模拟	易于实现求和亚阈值操作	难以实现高精度
数字	高精度和可编程性	难以实现好的 FF*
脉冲	中等精度,易于处理信号	难以实现好的 FF

* 填充因子。

本章的后半部分,介绍了除标准硅 CMOS 技术以外的其他几种 CMOS 图像传感器的结构和材料。近年来大规模集成电路的发展使得许多新的结构被提出,例如绝缘体上硅技术(silicon-on-insulator,SOI),蓝宝石上硅技术(silicon-on-sapphire,SOS),3D 集成以及 SiGe 和 Ge 等许多其他材料的使用。这些新结构和新材料的使用能够增强智能 CMOS 图像传感的性能和功能。

表 3.2 智能 CMOS 图像传感器的结构与材料

结构/材料	特 性
SOI	当使用 NMOS 和 PMOS 时面积小
SOS	透光性基底(蓝宝石)
3D 封装	大 FF
SiGe/Ge	长波长(NIR)

3.2 像素结构

本节,介绍了用于智能 CMOS 图像传感器的不同像素结构。虽然这些结构不同于传统的有源像素传感器,但是它们对于实现图像传感器的智能功能还是非常有用的。

3.2.1 电流模式

传统的 APS 把电压作为输出信号,但因为信号更容易通过基尔霍夫电流定律来实现加和减,所以电流模式更适用于信号处理。对于一个算术单元来说,用电流镜电路很容易实现乘法,它也可以通过电流镜比值大于 1 来实现光电流的倍增。但我们应注意到,这种方法引入了像素固定图像噪声。

电流模式中,用电流复制电路可以实现存储功能[157]。这里同时介绍了电流模式的 FPN 抑制和模数转换器[158]。电流模式分为两类:直接输出模式和累积模式。

3.2.1.1 直接输出模式

在直接输出模式中,光电流直接从光电探测器(光电二极管或光电晶体管)输出[159,160]。从光电二极管中产生的光电流通常使用倍增或者不倍增的电流镜来传输。一些早期的智能图像传感器使用光电晶体管通过电流镜来输出电流。如前文所述,电流镜的比值通常被用来放大输入电流,然而这种结构在低光照水平下灵敏度低,并且由于电流镜的失配导致 FPN 很大。基本电路结构如图 3.1 所示。

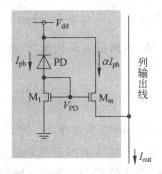

图 3.1 使用一个电流镜的像素基本电路结构(M_1 和 M_m 的宽长比是 α,所以电流镜的输出等于 αI_{ph})

3.2.1.2 累积模式

图 3.2 显示了电流模式有源像素传感器的基本结构[158,161,162]。采用有源像素传感器的结构与直接输出结构相比,图像质量有所提高。像素的输出可以用下式来表示

$$I_{pix} = g_m(V_{gs} - V_{th})^2 \tag{3.1}$$

其中,V_{gs} 和 g_m 分别是晶体管 M_{SF} 的栅源电压和跨导。在复位阶段,PD 节电压是

$$V_{reset} = \sqrt{\frac{2L_g}{\mu C_{ox} W_g}} I_{ref} + V_{th} \tag{3.2}$$

当有光照入射光电二极管时,PD 节电压变为

$$V_{PD} = V_{reset} - \Delta V \tag{3.3}$$

其中,T_{int} 是累加时间,ΔV 为

$$\Delta V = \frac{I_{ph} T_{int}}{C_{PD}} \tag{3.4}$$

这和电压模式 APS 的相同。因此输出电流可表示为

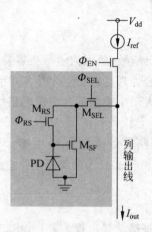

图 3.2 电流模式的 APS 的基本电路结构

$$I_{pix} = \frac{1}{2}\mu_n C_{ox} \frac{W_g}{L_g}(V_{reset} - \Delta V - V_{th})^2 \qquad (3.5)$$

于是差动电流 $I_{diff} = I_{ref} - I_{pix}$ 可以表示为

$$I_{diff} = \sqrt{2\mu_n C_{ox} \frac{W_g}{L_g} I_{ref}} \Delta V - \frac{1}{2}\mu_n C_{ox} \frac{W_g}{L_g} \Delta V^2 \qquad (3.6)$$

值得注意的是,由于晶体管 M_{SF} 的阈值电压被抵消,改善了源于阈值电压变化的 FPN。文献[161,162]中有更详细的介绍。

3.2.2 对数传感器

传统的图像传感器对输入光强的响应是线性的。对数传感器是一种基于 MOSFET 的亚阈值工作模式的传感器,亚阈值工作模式的解释见附录 E。对数传感器像素使用电流直接输出模式,因为当光电流小到使晶体管进入亚阈值区域时,电流镜的结构就是对数传感器结构。对数传感器的另外一种应用是宽动态范围图像传感器[163-168]。在 4.4 节中将会介绍宽动态范围图像传感器。

图 3.3 显示了对数 CMOS 图像传感器的基本像素电路结构。在亚阈值区域,MOSFET 的漏电流 I_d 非常小并随着栅电压 V_g 呈指数增长

$$I_d = I_o \exp\left(\frac{e}{mk_B T}(V_g - V_{th})\right) \qquad (3.7)$$

关于这个公式的推导和参数的意义,请查看附录 E。

在图 3.3(b)的对数传感器中,

$$V_G = \frac{mkT}{e}\ln\left(\frac{I_{ph}}{I_o}\right) + V_{ps} + V_{th} \qquad (3.8)$$

这种对数传感器结构采用了累积模式。M_C 的漏电流 I_c 可表示为

$$I_c = I_o \exp\left[\frac{e}{mk_B T}(V_G - V_{out} - V_{th})\right] \qquad (3.9)$$

由于电流 I_c 对电容 C 充电,从而 V_{out} 随时间的变化可表示为

$$C \frac{dV_{out}}{d_t} = I_c d t \qquad (3.10)$$

第 3 章 智能功能和材料

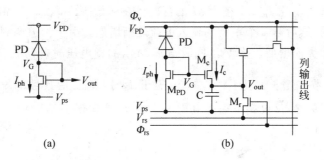

图 3.3 对数 CMOS 图像传感器的像素电路：(a) 基本的像素电路；(b) 包括累积模式的像素电路

将式(3.8)代入式(3.9)中，可得到

$$I_c = I_{ph} \exp\left[\frac{e}{mk_BT}(V_{out} - V_{ps})\right] \tag{3.11}$$

再将式(3.11)代入式(3.10)后积分，则输出电压 V_{out} 可表示为

$$V_{out} = \frac{mkT}{e} \ln\left(\frac{e}{mkTC}\int I_{ph} dt\right) + V_{ps} \tag{3.12}$$

尽管对数传感器具有可超过 100dB 的宽动态范围，但它也有一些缺陷，例如与 4T-APS 相比对数传感器光敏性差，尤其是在低光照条件下。此外，由于工作于亚阈值区使得响应缓慢，会造成比较大的器件特征偏差。

3.3 模拟域操作

本节介绍一些基本的模拟域操作。

3.3.1 WTA 模式

WTA(winner-take-all)电路是一种模拟电路[170]。图 3.4 所示为有 N 个电流输入的 WTA 电路，每个 WTA 单元由两个栅漏互通的 MOS 管 $M_{i(k)}$ 和 $M_{e(k)}$ 组成[171]。这里只考虑第 k 个和第 $(k+1)$ 个 WTA 单元。WTA 操作原理的关键特征是：晶体管 $M_{i(k)}$ 工作在具有沟道长度调制效应的饱和区并作为负反馈，而晶体管 $M_{e(k)}$ 工作在亚阈值区并作为正反馈。

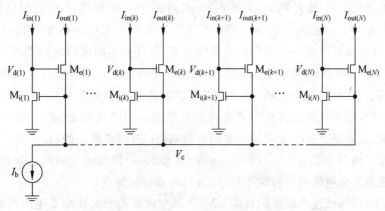

图 3.4 WTA 电路[171]

考虑输入电流 $I_{in(k)}$ 比 $I_{in(k+1)}$ 稍大 δI 的情况。在初始阶段,第 k 个和第 $k+1$ 个单元的初始电流是相同的,然后第 k 个单元的输入电流逐渐增加。由于沟道调制效应(见附录 E),$V_{d(k)}$ 随着 $I_{in(k)}$ 的增大而增加。当 $V_{d(k)}$ 同时也是 $M_{e(k)}$ 的栅电压时,$M_{e(k)}$ 工作在亚阈值区,其漏电流 $I_{out(k)}$ 呈指数式增长。当 $\sum_{i}^{N} I_{out(i)}$ 恒等于 I_b 时,$I_{out(k)}$ 指数性的增长引起其他电流 $I_{out(i)}(i \neq k)$ 减少。最终,仅仅剩下输入电流 $I_{in(k)}$ 流动,即实现了 WTA。

3.3.2 映射

映射是一种将像素值映射到一个方向的方法,如图 3.5 所示,通常是行和列的方向。映射将数据从 $M \times N$ 压缩到 $M+N$,这里 M 和 N 分别为行号和列号。映射操作既简单又快速,所以它在图像处理中是一种有效的预处理方法[172]。在电流模式中,把水平方向和垂直方向的输出电流相加就可以轻松实现映射操作。

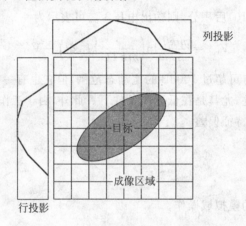

图 3.5 一个物体在行和列的方向上的映射

3.3.3 电阻网络

受到生物信号处理系统的启发,C. Mead 首先提出了硅电阻网络,这是一种实时超低功耗的大规模并行处理网络,详细的分析见文献[1]。

在这里先介绍一个硅视网膜的例子,硅视网膜是一种使用了电阻网络结构的智能 CMOS 图像传感器。在这种传感器中,MOSFET 作为电阻工作。就如同视网膜的边缘检测过程中,硅视网膜处理输入光图像的边缘和过零点,其基本的电路结构如图 3.6(a)所示。图 3.6(b)为一种一维网络结构的示意图,此处输入光的转化电压 $V_{i(k)}$ 被输入到网络中来分压,扩散电压 $V_{n(k)}$ 和输入光信号 $V_{i(k)}$ 输入到差分放大器中,$V_{o(k)}$ 是放大器的输出。如图 3.6(c)所示,电阻网络是输入光图像或模糊光图像的平滑器。如图 3.6(a)所示,它模仿了视网膜中的水平细胞,感光器是通过光电晶体管的对数响应(PTr)来实现的,在 3.2.2 节中已经讨论过它的功能。用差分放大器可以实现 on 和 off 单元,当图像输入的时候会自动进行边缘检测。需要注意二维电阻网络可能不稳定,所以需要仔细地设计。

文献[1]中提到,已经有很多的智能 CMOS 图像传感器研究使用了电阻网络。最近,带有噪声消除电路的三管有源像素传感器也已经应用到电阻网络[45,46]。图 3.7 展示了这种

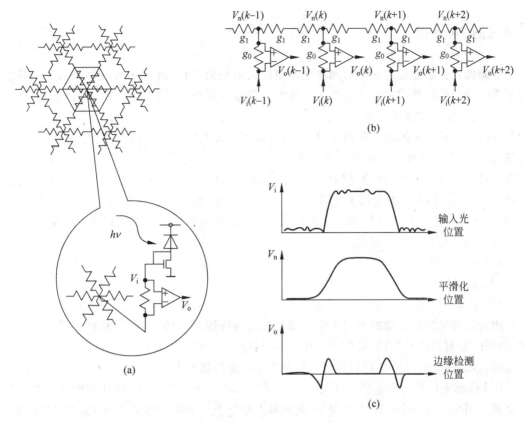

图 3.6 使用一个电阻网络的硅视网膜概念的图解:(a)电路图;(b)一维电路网络;(c)输入光模式和其他处理模式

传感器的基本结构,它包括两层网络。一种 100×100 像素的硅视网膜在高级图像处理应用中已经实现了商用化[46],这种传感器能够处理时域和空域的图像,所以应用在目标追踪等领域,这部分将在 4.7 节中作详细的讨论。

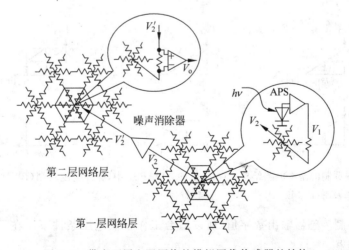

图 3.7 带有两层电阻网络的模拟图像传感器的结构

3.4 脉冲调制

有源像素传感器经过一定时间才能读出输出信号值,而在脉冲调制(PM)中,当信号达到了某一特定值时输出信号才会产生。这类使用脉冲调制的传感器叫做 PM 传感器、饱和时间传感器[176]或地址事件表示传感器[54]。脉冲宽度调制(PWM)和脉冲频率调制(PFM)的基本结构如图 3.8 所示。其他脉冲方法(如脉冲振幅调制和脉冲相位调制)很少用于智能传感器。K. P. Frohmader 最早提出了基于 PFM 的图像传感器的概念[177],K. Tanaka 等人发表了第一个基于 PFM 图像感知的应用[108],他使用 GAAS MSM 光探测器来演示该应用的基本工作情况,MSM 光探测器在 2.3.5 节作了讨论。基于 PWM 的图像传感器由 R. Muller 首次提出[178],而 V. Brajovic 和 T. Kanade 首次展示了它在传感器中的应用[179]。

脉冲调制有以下几个特点:

(1) 异步工作;

(2) 数字输出;

(3) 低电压工作。

因为脉冲调制传感器中的每个像素都能独立地决定是否输出,所以脉冲传感器可以在没有时钟(也就是说异步)的条件下工作。这个特点给 PM 图像传感器提供了对环境光照的自适应特性,因此可以考虑应用到宽动态范围的图像传感器中。

PM 传感器的另一个重要特征是其可以用作 ADC。在 PWM 中,脉冲宽度的计数值是数字值。图 3.8 是一个 PWM 的例子,其本质上相当于一个单斜坡型 ADC,而 PFM 相当于一种 1 位的 ADC。

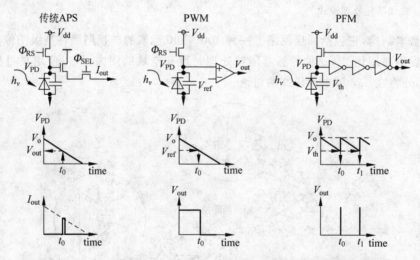

图 3.8 脉冲调制的基本电路结构:左,传统的 3T-APS;中,脉冲宽度调制(PWM);右,脉冲频率调制(PFM)

因为脉冲调制传感器输出数字值,所以适合工作在低电压条件下。在下一节将介绍几种脉冲调制传感器。

3.4.1 脉冲宽度调制

R. Muller 首次提出并实现了基于 PWM 的图像传感器[178]，后来 V. Brajovic 和 T. Kanade 提出并实现了一种使用 PWM 光电探测器的图像传感器。在这种传感器中，增加电路用来计算全局操作时处于导通状态的像素的数量总和，可以考虑用强度直方图获得累计演变。

这种数字输出方案适合芯片上的信号处理。M. Nagata 等人提出并证实了使用 PWM 的时域处理方案，并声明 PWM 适用于基于深亚微米技术的低电压和低功耗设计[180]，他们还展示了一种能够实现片上信号块平均化和二维映射的 PWM 图像传感器[181]。

文献[182,183]讨论了 PWM 的低电压工作特性，其中基于 PWM 的图像传感器在低于 1V 的电源电压下工作。特别是，S. Shishido 等人验证了一种由三个晶体管加上一个 PD 组成像素的 PWM 图像传感器[183]，这种设计克服了传统 PWM 图像传感器仅仅一个比较器就需要很多晶体管的缺点。

PWM 可以用来提高图像传感器的动态范围，这在 4.4.5 节中将详细论述，目前有很多关于这个课题的研究成果发表，文献[176]中讨论了 PWM 这种用途的几个优点，包括改善 DR 和 PWM 的信噪比。

PWM 也可以在数字图像传感器中用作像素级 ADC[57,60-62,67,140,184-186]，有些传感器为了尽量减少单位像素的面积而使用一个简单的反相器作为比较器，这样的话多个像素可以使用一个处理单元[56,67]。W. Bidermann 等人已经在一个芯片上实现了传统的比较器和存储器[186]。

图 3.9(b)中，一个斜坡信号被输入到比较器的参考端，这种电路几乎和单斜 ADC 相同，这种类型的 PWM 能够利用斜坡信号同步地运作。

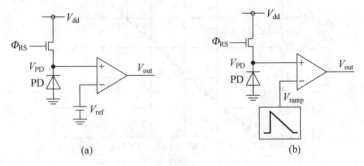

图 3.9 基于脉冲宽度调制(PWM)的光电传感器的基本电路结构，图中显示了两种类型的 PWM 光电传感器：一个是使用固定阈值的比较器(a)；另外一个使用斜坡信号的比较器(b)

3.4.2 脉冲频率调制

当累积信号达到阈值时，PWM 产生一个输出信号。同样在 PFM 中，累积信号达到阈值时产生输出信号，累积电荷重置时累计过程会重新开始，重复这一过程，会持续产生输出信号，输出信号产生的频率与输入光强度成正比。人们在生物系统中发现了类似于编码系统的 PFM[187]，它激励脉冲信号进行处理过程[188,189]，在第 5 章中将论述到 K. Kagawa 等人研究的脉冲式图像处理过程[190]。T. Hammadou 已经论证了 PFM 下的随机算法[191]。

K. P. Frohmader 等人首先提出并验证了基于 PFM 的感光传感器[177]，K. Tanaka 等人[108] 首先提出了基于 PFM 的图像传感器，W. Yang[192] 在宽动态范围成功验证了基于 PFM 的图像传感器，文献中有更多叙述[141,193,194]。

PFM 的一项应用是地址事件表达（AER）[195,196]，例如它可以用于传感器网络摄像机系统[197,198]。

PFM 感光传感器可以用于生物医学领域，例如极低光照下的生物技术探测[199,200]。另一项生物医学领域的应用是视觉假体，文献[201]中首次提出在视网膜下植入 PFM 感光传感器的视觉假体后，原小组[190,193,202-212]和其他一些小组[213-216]不断继续深入研究。5.4.2 节将论述视觉假体。

3.4.2.1 PFM 的工作原理

图 3.10 所示的是 PFM 感光传感器单元的基本电路，在电路中，C_{PD} 先被充电到 V_{dd}，然后包括暗电流 I_d 在内的总光电流 I_{ph} 使 PD 电容 C_{PD} 放电，结果引起 V_{PD} 减小。当 V_{PD} 达到反相器的阈值电压 V_{th} 时，反相器链开始导通，产生输出脉冲。输出频率 f 可以大致表达为

$$f \approx \frac{I_{ph}}{C_{PD}(V_{dd} - V_{th})} \quad (3.13)$$

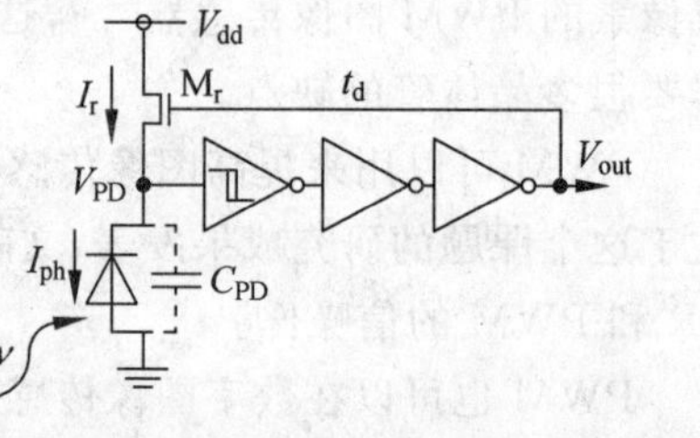

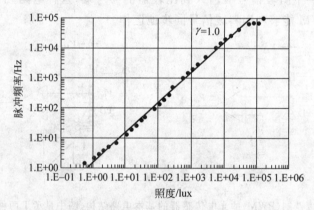

图 3.10 PFM 光传感器单元的基本电路

图 3.11 所示的是图 3.10 中 PFM 光传感器的实验结果，输出频率与输入光强度成正比，测量得到的动态范围接近 100dB，在低光强区域，饱和频率由暗电流产生。

图 3.11 图 3.10 所示 PFM 光传感器的实验输出脉冲频与输入光强度的关系

包含施密特触发器的反相器有一个延时键 t_d，由复位晶体管 M_r 提供的复位电流 I_r 有一个上限。以下是考虑这些参数在内的分析，更细致的分析见文献[206]，PD 通过光电流 I_{ph} 放电，因为光电流在电容充电过程中仍会产生，所以 PD 由复位电流 I_r 减去 I_{ph} 的电流充电。

考虑 t_d 和 I_r，V_{PD} 的变化如图 3.12 所示，V_{PD} 的最大电压 V_{max} 和最小电压 V_{min} 可以表示为

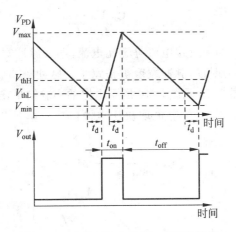

图 3.12 考虑 t_d 在内的 V_{PD} 的时间进程

$$V_{\max} = V_{thH} + \frac{t_d(I_r - I_{ph})}{C_{PD}} \tag{3.14}$$

$$V_{\min} = V_{thL} - \frac{t_d I_{ph}}{C_{PD}} \tag{3.15}$$

在这里，V_{thH} 和 V_{thL} 是施密特触发器的高阈值电压和低阈值电压，应该注意的是，I_{ph} 是放电电流，而充电电流或复位电流是 $I_r - I_{ph}$。图 3.12 给出的 t_{on} 和 t_{off} 的值为

$$t_{on} = \frac{C_{PD}(V_{thH} - V_{\min})}{I_r - I_{ph}} + t_d = \frac{C_{PD}V_{th} + t_d I_r}{I_r - I_{ph}} \tag{3.16}$$

$$t_{off} = \frac{C_{PD}(V_{\max} - V_{thL})}{I_{ph}} + t_d = \frac{C_{PD}V_{th} + t_d I_r}{I_{ph}} \tag{3.17}$$

其中，$V_{th} = V_{thH} - V_{thL}$，$t_{on}$ 是复位晶体管 M_r 给 PD 充电的时间，即 M_r 开启的时间，在这段时间内，脉冲信号为高电平，因此它等于脉冲宽度。t_{off} 是 M_r 断开的时间，在这段时间内，脉冲信号为低电平。PFM 光传感器的脉冲频率 f 表示为

$$\begin{aligned} f &= \frac{1}{t_{on} + t_{off}} \\ &= \frac{I_{ph}(I_r - I_{ph})}{I_r(C_{PD}V_{th} + t_d I_r)} \\ &= \frac{I_r^2/4 - (I_{ph} - I_r/2)^2}{I_r(C_{PD}V_{th} + t_d I_r)} \end{aligned} \tag{3.18}$$

如果 M_r 的复位电流 I_r 远大于光电流 I_{ph}，那么式(3.18)就变成

$$f \approx \frac{I_{ph}}{C_{PD}V_{th} + t_d I_r} \tag{3.19}$$

从而脉冲频率 f 就与光电流 I_{ph} 成正比，也就是与输入光强成正比。

另外，由式(3.18)可以看出频率 f 在光电流是 $I_r/2$ 时达到最大值，然后开始下降，其最大频率 $f_{\max}$ 是

$$f_{\max} = \frac{I_r}{4(C_{PD} + t_d I_r)} \tag{3.20}$$

脉冲宽度 τ 是

$$\tau = t_{on} = \frac{C_{PD}V_{th} + t_d I_r}{I_r - I_{ph}} \tag{3.21}$$

由式(3.21)可以看出,如果复位电流等于光电流,那么脉冲宽度变为无穷大,即当输入光强度强或者复位电流小的时候,脉冲宽度被拓宽了。脉冲频率和脉冲宽度与输入光强度的关系如图3.13所示。图3.14是图3.10所示电路的实验结果,当$V_{dd}=0.7V$并且输入光强度比较大时,脉冲宽度变宽,复位电流取决于电源电压V_{dd},所以脉冲宽度变宽的结果是合理的。

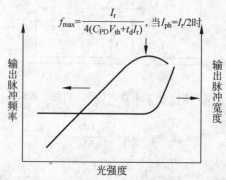

图3.13 脉冲频率和脉冲宽度取决于输入光强度

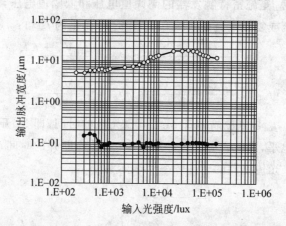

图3.14 图3.10所示电路由输入光强度决定的脉冲宽度的实验结果

3.4.2.2 电容反馈PFM

上述由复位电流和光电流引起的竞争效应可以通过引入电容反馈来减缓[54,193]。图3.15所示的是改进过的电容反馈的电路原理图,在电容反馈PFM光传感器中,当V_{PD}接近INV1的阈值电压时,INV2的输出状态逐渐由LO变为HI。输出电压通过复位晶体管M_{rst}中的重叠电容C_{rst}引起V_{PD}的正反馈,从而加速V_{PD}的下降。正反馈带来几大好处,最明显的是复位电流和光电流之间的竞争显著减小,所以电源电压就可以在没有任何脉冲特征退化的情况下减小。另外,因为由一个反相器链产生的正反馈不产生延时,所以反相器的个数可以减少。在我们的实验中,反相器的两个阶段就足以正常运转。在图3.16是由$0.35\mu m$标准CMOS工艺制成的电容反馈感光传感器的实验结果[209],甚至在电源电压不足1V时,也有令人满意的实验结果。

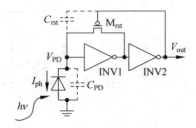

图 3.15 带有电容反馈复位的 PFM 像素的原理图

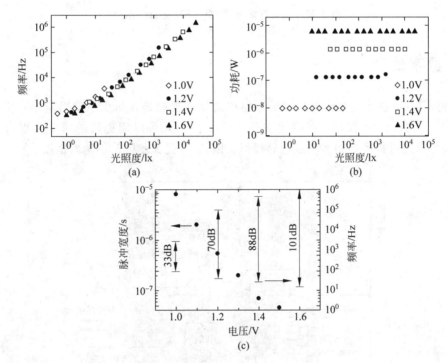

图 3.16 (a)输出脉冲频率相关性的实验结果;(b)光照的能量损耗;(c)电容反馈 PFM 光传感器的脉冲宽度和饱和频率与电源电压的关系。这个光传感器基于 $0.35\mu m$ 标准 CMOS 工艺,像素面积是 $100\mu m \times 100\mu m$,光二极管尺寸是 $7.75\mu m \times 15.60\mu m$[209]

3.4.2.3 带有恒定偏压 PD 的 PFM

以上所述的 PFM 结构都会改变 PD 的电压,它随输入光强或光电流而线性变化。图 3.17 是一个带有恒定 PD 偏压的 PFM 像素电路[199],PD 的负极电压实际是运算放大器(OPamp)的正输入端电压,所以该节点电压固定在 V_{ref}。反馈电容 C_{int} 的存储电荷被光电流抽走,因此 OPamp 的输出节点电压升高。当 OPamp 的输出电压达到比较器(Comp)的阈值电压时,Comp 导通,复位晶体管 M_{rst} 被比较器的输出脉冲关断。

Comp 的输出脉冲的频率与输入光强成正比。这种 PFM 像素电路最重要的特点是 PD 的偏压是一个常数,可以通过降低 PD 偏压来抑制其自身的暗电流。它可以应用于低光强度下的检测,这将在 4.2 节讨论。

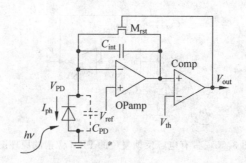

图 3.17 带有常偏压 PD 的 PFM 像素电路[199]

3.5 数字处理

智能 CMOS 图像传感器的数字处理结构的设计思想是在每个像素里应用一个数字处理单元[44,57,58,61,62,67]。在每个像素里应用数字处理单元可以用更快的速度实现可编程操作,并且可以通过数字结构实现最近邻操作。图 3.18 是文献[63]的像素框图。

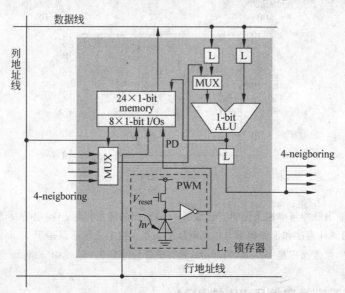

图 3.18 具有数字处理结构的智能 CMOS 图像传感器的像素电路图

数字处理结构的关键是在每个像素中应用一个 ADC。文献[44]介绍了阈值的二元运算。文献[57,67]介绍了一种用反相器来实现比较器的简单 PWM 方案。因为每个像素都会传输数字信号到下一个像素,所以使用这个传感器不需要扫描电路,这是完全可编程数字处理架构的另一个功能。在文献[217]中,像素级 ADC 可以有效地控制转换曲线和伽马值。传感器的可编程性用来加强捕获的图像,如在对数转换、直方图均衡化以及类似的技术。

全数字处理结构非常适合于智能 CMOS 图像传感器,它可以通过编程得到很高的精度,很适合于机器人视觉,因为机器人视觉需要通过快速响应作出全方位的自主操作。数字处理技术的挑战在于像素分辨率,这是由于一个像素内包含太多晶体管会受到制约,例如在文献[61]中,应用 0.5μm 标准 CMOS 工艺技术,一个 80μm×80μm 的像素中集成了 84 个晶体管。通过使用小尺寸的 CMOS 工艺,可以制成面积更小、能耗更低且处理速度更快的像素。但是,这样的工艺面临低电压摆幅和低感光灵敏度等方面的难题。

另外一种应用于智能 CMOS 图像传感器的数字处理技术是,在数字像素传感器中由 4 个像素共用一个 ADC[140,185,186],这种数字像素传感器达到 10000fps[185] 的高速度和超过 100dB 的宽动态范围[186]。

3.6 硅以外的材料

这一节将会介绍可用于智能 CMOS 图像传感器的除硅以外的几种材料。通过第 2 章的介绍可知,硅可以吸收可见光,也就是说硅对于可见光波长是不透明的。有一些材料在可见光波长区域是透明的,这在现代 CMOS 工艺中(如 SOI 和 SOS)都有用到,比如 SiO_2 和 Al_2O_3。硅的可检测波长取决于它的能带隙,大约是 $1.1\mu m$ 的波长。与硅相比,其他材料(如 SiGe 和锗)能够响应波长更长的光,详述见附录 A。

3.6.1 绝缘体上硅

近年,SOI CMOS 工艺已经被应用于低电压电路[218]。这种 SOI 的结构如图 3.19 所示,一层薄硅层位于一层氧化物掩埋层(BOX)上,顶层硅位于 SiO_2 层或绝缘层上。传统的 CMOS 晶体管被称作体 MOS 晶体管,这样可以明显地区别于 SOI MOS 晶体管。MOS 晶体管制作在一层 SOI 上并通过进入到 BOX 层的浅沟槽隔离工艺(STI)被完全隔离,见图 3.19(b)。相比于体 MOS 晶体管,这种晶体管具有较低能耗、较小的闩锁效应和较低的寄生电容,以及其他的一些优点[218]。

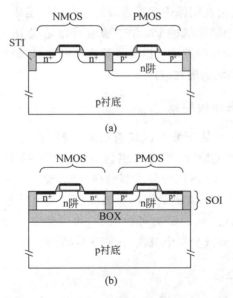

图 3.19 (a)体 CMOS 的横截面;(b)SOI CMOS 的横截面。STI:浅沟槽隔离,SOI:绝缘体上硅,BOX:掩埋氧化层。硅化物是由硅和金属(如 TiSi2)制成的合成材料

SOI 工艺适合于 CMOS 图像传感器的原因如下：
- 利用 SOI 工艺制成的电路电压低、功耗小，这种特性对于移动设备、传感器网络、可植入医疗设备来说非常重要。
- 体 CMOS 晶体管工艺中的 n 阱层对于在 p 型衬底上制作的 PMOS 是必不可少的，而当使用 SOI 工艺时，NMOS 和 PMOS 晶体管都可以避免牺牲面积。图 3.19 给出了体工艺和 SOI 工艺的对比图。与 NMOSFET 复位晶体管相比，PMOSFET 复位晶体管由于没有电压降更适用于 APS 中。
- SOI 技术使制造背照式图像传感器更加容易，这会在后续章节中介绍。
- SOI 结构有利于防止像素之间[220]由于扩散载流子而引起的串扰，衬底产生的光生载流子可以到达 SOI 图像传感器的像素当中。在 SOI 技术中每个像素都是电学隔离的。
- SOI 工艺也可以用于三维集成[221,222]，文献[221]将论述在图像传感器中使用 SOI 技术实现一种首创性的三维集成研究。

将 SOI 工艺用于 CMOS 图像传感器的难题之一就是如何实现光检测。通常 SOI 层很薄（一般小于 200nm）以至于光灵敏度会降低，为了实现高的光灵敏度，人们提出了几种解决方法。一种与传统 APS 最兼容的方法是在衬底上制作一个 PD 区域[223-226]，这就保证了其光灵敏度与传统 PD 的光灵敏度一样，但是这需要改变标准 SOI 工艺的制作方法，表面后处理的工艺对于获得低的暗电流也很重要。第二种方法是使用一个横向 PTr，在 2.3 节[227-230]的图 2.7(d)有介绍。因为横向 PTr 有增益，所以即使在很薄的光检测层中，光灵敏度也会增加。SOI 的另一种应用在于横向 pin PD，尽管在这种情况下光探测区域和像素密度需要权衡考虑。在 3.4.2 节中介绍的 PFM 光电传感器非常适合 SOS 成像器，本章中还会有进一步讨论。

通过有选择性地使用 SiO_2 刻蚀，BOX 层可以轻松得到一束硅结构，所以 SOI 被广泛应用于微机电系统。这种结构在图像传感器领域的一个应用是非冷却的红外焦平面阵列图像传感器(FPA)[232]。用一个热绝缘的 PN 结二极管可以完成 IR 检测，热辐射使 PN 结内建电势差发生变化，通过感知这种波动可以测量温度和检测 IR 辐射。结合 MEMS 的结构，拓宽了 SOI 在图像传感器领域的潜在应用。

3.6.1.1 背照式图像传感器

在这里我们会介绍在可见光波长区域透明的材料，它适合于背照式图像传感器。由图 3.20 可以看出，背照式 CMOS 图像传感器具有大 FF 值和大光学反应角度的优势。图 3.20(a)显示了传统的 CMOS 图像传感器的横截面，输入光从微型镜面到 PD 要经过很长的距离，这引起了像素之间的串扰，此外，金属线也是光的阻碍物。在背照式 CMOS 图像传感器中，微型镜面到 PD 之间的距离可以减少，得以使其光学特性大大提高。PD 上的 p-Si 层必须要薄到以尽可能减少层中吸收，衬底也尽量得很薄。

3.6.1.2 蓝宝石上硅

蓝宝石上硅(SOS)是使用蓝宝石代替硅作为衬底的一种技术，它将一层薄硅直接生长在蓝宝石衬底上。应该注意的是，顶层硅不是多晶硅或非晶硅而是单一的晶体硅，因此它的流动性等物理属性几乎和一个普通的 Si-MOSFET 一样。蓝宝石是氧化铝，它对于可见波

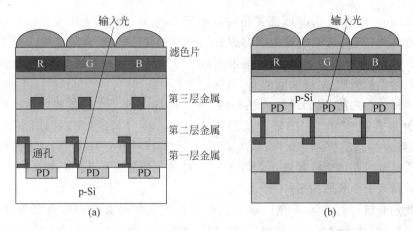

图 3.20 (a)一个传统 CMOS 图像传感器的横截面；(b)一个背照式 CMOS 图像传感器的横截面

长区域是透明的,所以使用 SOS 技术的图像传感器不通过任何打薄处理就可以用作背照式图像传感器[203,231,234,235],但是为了获得平整的背面,实际上还是需要一些打磨。横向 PTr 在文献[231,235]中有提到,PFM 光电传感器因为在薄的探测层中有低的光灵敏度在文献[203,234]有提到。图 3.21 是一个用 SOS CMOS 技术制成的图像传感器,这个芯片被放在一叠打印出来的纸上,可以透过透明的衬底清楚看到纸上的打印图案。

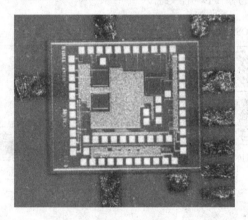

图 3.21 应用 SOS 技术制成的 PFM 光电传感器[203]

3.6.2 扩展检测波长

通常硅的灵敏度能高达到 $1.1\mu m$,这是由硅的能带隙 $E_g(Si)=1.12eV$ 决定的。为了扩展灵敏度到 $1.1\mu m$ 以上,需要用 Si 以外的其他材料,有许多材料的灵敏度在比硅更长的波长范围内。为了实现灵敏度的适用波长范围比 Si 更长的智能 CMOS 图像传感器,需要一些在更长波长范围有良好光响应率的混合集成材料,如 SiGe、Ge、HgCdTe、InSb 和量子阱红外线光电探测器(QWIP)[236]以及一些其他的材料[237],除了 SiGe 外,这些材料都可以被放置在一个通过倒装芯片结合金属凸块组成的硅读出集成电路(ROIC)上。几种使用 ROIC 来实现 IR 检测的方法见文献[238]。肖特基势垒光电探测器如 PtSi(硅化铂)也广泛应用于 IR 成像系统[239],它可单片集成在硅衬底上。通常这些红外图像传感器工作在冷却条件下,现在已经有许多图像传感器具有这种结构,它超出了本书的范围,我们不再介绍。

在这里,只介绍一个智能 CMOS 图像传感器的例子,它在被称为人眼安全波长区域的可见光区域和近红外(NIR)区域的灵敏度都很好。相对于可见区域,人眼睛对人眼安全波长区域($1.4\sim2.0\mu m$)有更高的忍耐度,这是因为对于可见光区域的光线,人眼安全波长区域的光更容易在角膜被吸收,对视网膜的损害更小。

在描述传感器之前,我们简要描述材料 SiGe 和 Ge。Si_xGe_{1-x} 是硅和锗以任意比例 x 混合的一种混合晶体[240],它的能带间隙可以从 $Si(x=1)$,$E_g(Si)=1.12\,eV$,$\lambda_g(Si)=1.1\mu m$ 到锗 $(x=0)$,$E_g(Ge)=0.66\,eV$,$\lambda_g(Ge)=1.88\mu m$ 变化。硅上的 SiGe 可以用于高速电路的异质结双极晶体管或者应变 MOSFET。硅和锗的晶格常数之间的晶格失配很大,导致很难在硅衬底上生长厚的 SiGe 外延层。

现在我们介绍一种在可见光区域和人眼安全波长区域都可以工作的智能图像传感器[241,242]。这种传感器包含一个传统 Si CMOS 图像传感器和一个位于 CMOS 图像传感器下面的锗 PD 阵列,这种传感器捕获可见图像的能力并不会由于其范围扩大到 IR 区域而受影响。NIR 探测的工作原理是基于从锗 PD 注入到 Si 衬底的光生载流子,这种结构如图 3.22 所示,在锗 PD 区域产生的光生载流子注入到 Si 衬底中,然后通过扩散到达 CMOS 图像传感器像素中的光转化区域。如图 3.23 所示,当施加偏压时,NIR 区域的响应率增加。图中的插图是光敏反应实验的测试装置。因为在 NIR 波长区域 Si 衬底是透明的,所以 NIR 光可以被位于传感器背面的锗 PD 检测到。

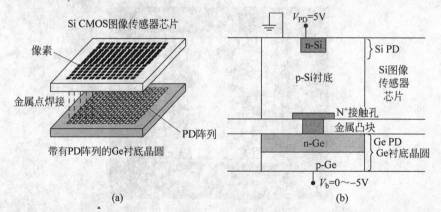

图 3.22 可以在可见光区域和人眼安全区域都检测的智能传感器:(a)芯片结构;(b)传感器横截面[242]

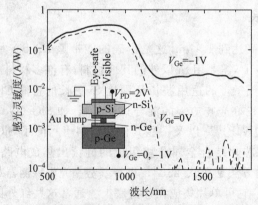

图 3.23 随输入光波长变化的光灵敏度变化曲线,锗 PD 的偏压 V_b 是一个参量。插图是光敏反应实验的测试装置

3.7 非标准 CMOS 技术结构

3.7.1 3D 集成

三维(3D)集成可以在有限区域集成更多的电路[243,244]。层之间的互连通过微通孔[221,222,243,221]、电感耦合[245,246]、电容耦合[247]和光学耦合[248]来实现。一些传感器使用 SOI[221,222,243]和 SOS[247]技术,这使得两个晶片之间的键合变得更加容易。

由于生物眼有垂直分层结构,3D 集成可以用于模仿生物系统[249,250]。一个具有 3D 集成结构的图像传感器在它的上表面有成像区域,在它的连续层有信号处理电路,所以 3D 集成技术使像素级处理或像素并行处理变得更简单。图 3.24 是 3D 图像传感器芯片的概念图及其横截面结构图[249]。在 5.4.2 节中将要讨论 3D 图像传感器作为一个视网膜假体设备的功能[213],3D 图像传感器因其像素的并行处理功能而大有前途,然而进一步的发展要求它具有与传统 2D 图像传感器相媲美的图像质量。

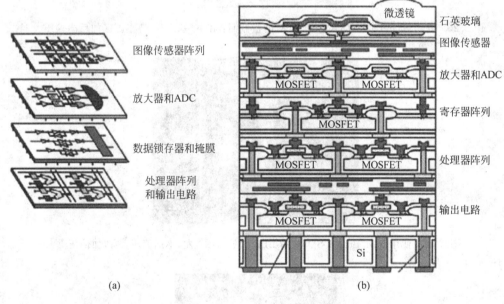

图 3.24 3D 图像传感器芯片:(a)结构;(b)横截面结构[249]

3.7.2 集成光发射器

CMOS 图像传感器的光源集成将会开发出来许多应用,比如超小相机系统和自动机器人视觉。Ⅲ-Ⅴ族化合物半导体,如 GaAs[86]、纳米晶体硅、铒掺杂硅[251]和 β-FeSi$_2$ 都有很好的光发射效率,但是它们与标准 CMOS 工艺难以兼容。另外,铒掺杂硅和 β-FeSi$_2$ 的发射波长比 Si 的带隙波长要长。虽然硅有间接能带,但它可以通过带间发射方式发射光[252],具有 1‰的良好的发射效率,也可以通过发射效率更低的热电子发射[253-257]。给 PN 结施加正向偏压,就会发生带间发射。发射峰约为 1.16μm,这是由硅的带隙能量决定的,因此它不能作为 CMOS 图像传感器的光源,它在这个波长的光灵敏度很低。PN 结二极管反向偏置时,二极管会发射中心波长约为 700nm 的宽光谱光（超过 200nm），它可以由硅 PD 检测到，这种宽谱发射来源于雪崩击穿时的热电子[258,259]。

文献[255]中演示了一个集成 Si-LED 基于标准 SiGe-BiCMOS 工艺的图像传感器。如图 3.25 所示，使用 SiGe-BiCMOS 工艺是因为，p^+n^+ 二极管可以通过 p^+ 基区和 n^+ 沉降区中间的结得到，只是这种方法仅适用于 SiGe-BiCMOS 工艺。值得注意的是，发射不是来源于 SiGe 基区而是来源于 Si 区。SiGe-BiCMOS 电路的光发射如图 3.26 所示，包含 Si-LED 阵列的一种 CMOS 图像传感器已经研制成功，如图 3.27 所示，目前，集成在图像传感器上的 Si-LED 能耗非常高且发射效率低，通过优化结构应该可以改进这些特性。

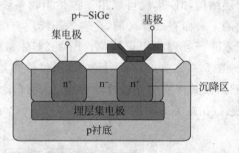

图 3.25　使用 SiGe-BiCMOS 工艺的 LED 的横截面：在 p^+ 基区和 n^+ 集电区加反向偏压[255]

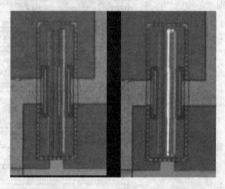

图 3.26　使用 SiGe-BiCMOS 工艺的 LED 的放射图：左，不加偏压；右，加偏压[255]

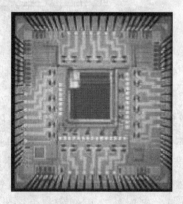

图 3.27　集成了 Si-LED 的图像传感器[255]

3.7.3 通过非标准结构实现颜色识别

图像传感器可以检测颜色信号,并将光分解成基本的颜色信号,如 RGB。传统的颜色识别的方法将在 5.4.1 节介绍。其他使用智能功能实现颜色识别的方法将在接下来的几个部分作总结。

3.7.3.1 堆叠有机 PC 薄膜

首先介绍一种可以获得 RGB 颜色的方法,即在一个像素上使用三个堆叠的光电导(PC)有机薄膜[104-106],每一个有机薄膜都相当于一个 PC 检测器(见 2.3.5 节),根据其感光性产生相应光电流,这种方法几乎可以实现 100% 的 FF。现在面临的主要问题是如何连接堆叠层。

3.7.3.2 多重结

硅的光敏性取决于 PN 结的结深。因此,位于一条垂直线的两个或三个结可以改变光敏范围[260-262]。调整三个结深度,相对应的 RGB 颜色的最大光敏性就可以实现。图 3.28 就是一个这种结构的传感器,做一个三重阱来形成三个不同的二极管[261,262],这个传感器作为 APS 类型像素已经商业化了。

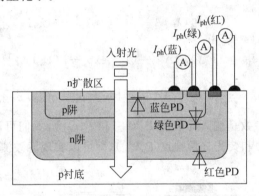

图 3.28 具有三结的图像传感器的结构[262]

3.7.3.3 控制电势分布

许多研究人员已经提出并证明了通过控制电势分布可以改变光谱灵敏度[263-265],这种系统主要使用由多层 p-i-i-n[263,264] 和 n-i-p-i-n[265] 构成的薄膜晶体管(TFT)层。Y. Maruyama 等人在丰桥大学提出了使用这种方法[266,267]的智能 CMOS 图像传感器,虽然他们的目的不是颜色识别,而是滤波荧光检测,这将在随后的章节讨论。

电势控制原理如下[266,267]。在 2.3.1.2 节中 PN 结 PD 的灵敏度通常由式(2.19)表示。这里的电势分布图如图 3.29 所示,该图是由图 2.11 变化来的。即通过用 n 衬底上的 PMOS 型 PG 代替 NMOS 型 PG,给出两个耗尽区,一个源于 PG,一个源于 PN 结。这个 PN 结产生一个凸出的电势,就像一个光生载流子的分水岭。此例中,式(2.18)的积分区域从 0 到 x_c 变化,因为这个变化范围内光生载流子到达表面和衬底的几率是相同的。移动

到衬底的光生载流子只引起光电流,所以灵敏度就变成了

$$R_{ph} = \eta_Q \frac{e\lambda}{hc}$$

$$= \frac{e\lambda}{hc} \frac{\int_0^{x_c} \alpha(\lambda) P_o \exp[-\alpha(\lambda)x] dx}{\int_0^{\infty} \alpha(\lambda) P_o \exp[-\alpha(\lambda)x] dx}$$

$$= \frac{e\lambda}{hc}(1 - \exp[-\alpha(\lambda)x_c] dx) \tag{3.22}$$

从中得出,如果有两束不同波长的光,激励光源 λ_{ex} 和荧光 λ_{fl} 同时入射,那么总的光电流 I_{ph} 就是

$$I_{ph} = P_o(\lambda_{ex}) A \frac{e\lambda_{ex}}{hc}(1 - \exp[-\alpha(\lambda_{ex})x_c])$$

$$+ P_o(\lambda_{fl}) A \frac{e\lambda_{fl}}{hc}(1 - \exp[-\alpha(\lambda_{fl})x_c]) \tag{3.23}$$

这里 $P_o(\lambda)$ 和 A 分别是 λ 的入射光功率密度和光栅(PG)面积。当有两种不同栅压计算光电流时,x_c 有两个不同的值 x_{c1} 和 x_{c2},这就导致有两个不同光电流 I_{ph1} 和 I_{ph2}:

$$\begin{cases} I_{ph1} = P_o(\lambda_{ex}) A \frac{e\lambda_{ex}}{hc}(1 - \exp[-\alpha(\lambda_{ex})x_{c1}]) \\ \qquad\quad + P_o(\lambda_{fl}) A \frac{e\lambda_{fl}}{hc}(1 - \exp[-\alpha(\lambda_{fl})x_{c1}]) \\ I_{ph2} = P_o(\lambda_{ex}) A \frac{e\lambda_{ex}}{hc}(1 - \exp[-\alpha(\lambda_{ex})x_{c2}]) \\ \qquad\quad + P_o(\lambda_{fl}) A \frac{e\lambda_{fl}}{hc}(1 - \exp[-\alpha(\lambda_{fl})x_{c1}]) \end{cases} \tag{3.24}$$

在这两个方程中,未知参数是输入光强度 $P_o(\lambda_{ex})$ 和 $P_o(\lambda_{fl})$,我们可以计算两个输入光强度即激发光源强度 $P_o(\lambda_{ex})$ 和荧光光源强度 $P_o(\lambda_{fl})$,即可以实现滤波测量。

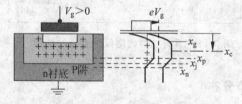

图 3.29 滤波荧光图像传感器的器件结构和电势分布图[266,267]

3.7.3.4 亚波长结构

实现颜色检测的第 4 种方法是使用亚波长结构,例如一个金属网格或表面等离子激元[268-270]和光子晶体[271],这些技术还处于起步阶段,但对于微间距像素的 CMOS 图像传感器可能是有效的。在亚波长结构,量子效率对极化、入射光波长和金属网格的形状和材料都非常敏感,这意味着光必须被视为电磁波来估计它的量子效率。

当孔径 d 比入射光波长 λ 小得多时,通过孔 T/f 的光传输随着 $(d/\lambda)^{4[272]}$ 而减少,这就会造成图像传感器的灵敏度以指数式降低,T/f 即在孔面积 f 上传输光强度 T 归一化入射

光的强度。T.Thio 等人提出了一种传输增强装置,它是通过在金属表面上周期性凹槽包围的亚波长孔径实现[273]。在这样的结构中,表面等离子体(SP)模式由入射光的光栅耦合[274]激发,SP 共振造成了通过孔径的光传输的增强,这种传输增强使得具有亚波长孔径的图像传感器的实现成为可能。通过文献[269]给出的计算机仿真结果可以看出,Al 金属网格增强了光传输,而钨金属网格却没有增强,金属网格的厚度和布线和距离也会影响到传输。

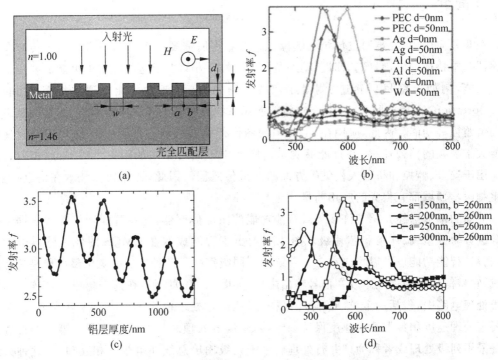

图 3.30 金属网格作为亚波长结构和它的光传播。计算机仿真(FDTD:时域有限差分)模型,(a)光传播与材料参数和波长的关系;(b)金属厚度;(c)周期;(d)PEC:理想电导体[269]

第 4 章 智能成像

4.1 简介

在某些应用中,很难通过传统的图像传感器获得所需要的图像。一方面由于传统图像传感器基本特征的限制,例如速度、动态范围等;另一方面由于这些应用中可能需要一些特殊的功能,例如追踪目标轨迹、测量距离等。例如,在将来的智能交通系统(intelligent transportation systems,ITSs)中,要求智能照相系统具有辅助车辆保持车道、测量距离以及驾驶员监控等功能。因此,应用在 ITSs 中的图像传感器的动态范围应该大于 100dB,速度也要大于视频的帧频速率,并且能够测量出图像中不同对象间的距离[275]。与此相似的还有应用于安全、监控和机器人视觉等方面的图像传感器。因此,智能成像技术在信息、通信和生物医学领域都有很大的应用前景。

前面的章节已经介绍了许多集成智能功能的图像传感器,这些智能功能通常依据其集成等级分为像素级、列级和芯片级。图 4.1 列出了智能成像在 CMOS 图像传感器中的分类,当然,智能功能也可以集成在一个系统的不同级别中。几种集成分类中最简单是芯片级集成,它将信号的处理电路放在信号输出之后,如图 4.1(a)所示,在这个集成图像传感器的信号处理电路中,包括一个 ADC、降噪系统、彩色信号处理模块等。应当注意的是,这种方法要求处理速度和数据读出速度保持一致,因此信号处理芯片的增益有限。第二种集成方式是采用列级处理或者称为列并行处理。由于列级输出总线在电学上相互独立,这种方式比较适合 CMOS 图像传感器。列级集成中,每一列都进行信号处理,因此其对信号处理速度的要求比芯片级集成要小,同时,列级集成可以采用传统 CMOS 图像传感器的像素结构,如 4T-APS 等,这使其在 SNR 方面有很大的优势。第三种集成方法是像素级集成,或者称为像素并行集成。在这种方法中,每一个像素拥有一个独立的信号处理电路和一个光电探测器,因此,可以实现快速且通用的信号处理。由于像素中加入了信号处理电路,其光电探测器的面积(或者说填充因子)会比前两种集成方式的要低,这会导致图像质量变差。同时,这种方法也很难将 4T-APS 应用于像素中。尽管有这些问题,这种结构始终还是有它的优势,它依然是下一代智能图像传感器的一个发展方向。

智能应用中的 CMOS 图像传感器的其他特性如下:

(1)任意像素的随机采集。

(2)非破坏性读出或并行读出。需要注意的是,原始的 4T-APS 不能进行非破坏性读出。

(3)在每一个像素和/或每一列和/或芯片上集成信号处理电路。

在本章中,我们将研究智能 CMOS 图像传感器的几个应用的智能成像要求。

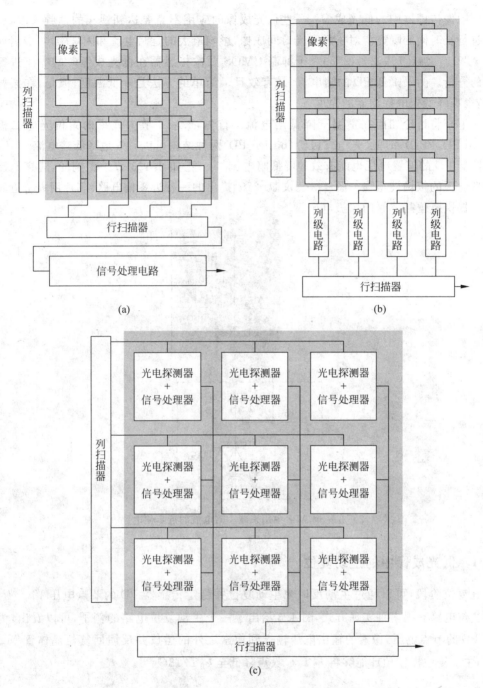

图 4.1 智能图像传感器的基本分类：(a)芯片级；(b)列级；(c)像素级

4.2 低光成像

在天文和生物技术等领域中，需要应用低光成像技术。某些系统中的图像传感器要求有超高的灵敏度，如超高空侦查站[102]和碰撞系统[276,277]，但在本节中我们的研究主要集中在智能图像传感器中的低光成像。一些低光成像的应用不需要达到视频的帧频速率,所以可以进行长时间的曝光。对于长曝光时间成像,影响最大的是暗电流和闪烁噪声(或者叫做 $1/f$ 噪声)。文献[199,200,278]对于抑制 CMOS 图像传感器在低光成像时的噪声进行了详细的分析。为了降低 PD 的暗电流,最有效且最简单的方法是降低温度,但是,在某些应用中,很难降低光电探测器的温度。

接下来将讨论如何在室温下降低暗电流。首先,2.4.3 节中介绍的钳位光电二极管(pinned PD,PPD)和埋层光电二极管(buried PD)两种结构可以有效地降低暗电流。

降低 PD 的偏置电压也能有效地降低暗电流[78]。正如 2.4.3 节中提到的,隧穿电流对于偏置电压的依赖性很强。图 4.2 为文献[279]提出的一种近零偏电路,芯片用一个近零偏电路来提供复位晶体管的栅极电压。

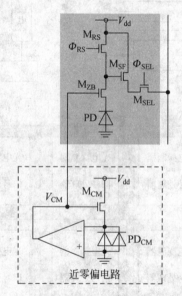

图 4.2　在一个 APS 中用来减小暗电流的近零偏电路[279]

4.2.1　低光成像中的主动复位

在复位阶段,可以通过主动反馈来主动复位并稳定传感器 PD 的节点电压 V_{PD}[280-288]。主动复位电路有很多种实现方式,图 4.3 给出了一个主动复位电路的例子,可以看出,为了稳定 PD 的节点电压,像素的输出电压输入到像素之外的运放并反馈回复位晶体管 M_{rst} 的栅极,主动复位操作预计能够将 $k_B T/C$ 噪声降低至 $k_B T/18C$[280]。

4.2.2　PFM 在低光成像中的应用

在 3.4.2.3 节中讨论过,带有近零偏 PD 的 PFM 图像传感器能够实现超低光探测,其最小探测信号大小为 0.15fA(积分时间为 1510s)。正如 2.3.1.3 节中提到的,PD 的暗电流与 PD 的偏置电压成指数关系,因此接近零伏的 PD 偏置电压能够有效地减少暗电流。在这种情况下,应该更多考虑其他的漏电流。对于一个 PD 偏置固定的 PFM 图像传感器来说,其复位晶体管 M_{rst}(见图 3.17)的漏电流和亚阈值电流对于信号的影响很大,应该尽可能减

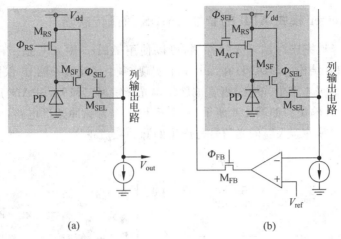

图 4.3 主动复位电路：(a)传统的 3T-APS；(b)带主动复位的 3T-APS[280]

小。图 4.4 为 Bolton 等人提出的一种降低亚阈值电流的方法。电路中复位晶体管的源漏电压接近于零，所以亚阈值电流接近于零。附录 E 中描述了亚阈值电流与源漏电压和栅源电压成指数关系的原因。

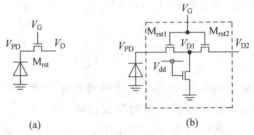

图 4.4 带固定 PD 偏置的 PFM 图像传感器的复位电路：(a)一个复位晶体管的电路；
(b)三个复位晶体管的电路

4.2.3 差分 APS

为了抑制共模噪声，文献[278]提出了差分的 APS，其结构如图 4.5 所示。传感器中采用了一个钳位 PD，并用 PMOS 电流源来完成低偏置操作。应当注意的是，PMOS 的 $1/f$ 噪声是小于 NMOS 的 $1/f$ 噪声的。最终，该传感器在室温下，经过超过 30s 的积分时间，完成了 10^{-6} 照度的超低光探测。

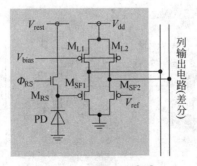

图 4.5 差分 APS[278]

4.2.4 采用 Geiger 模式 APD 的智能 CMOS 图像传感器

为了能够迅速地完成对超低光的探测,可以使用 APD。图 4.6 是一种在标准 CMOS 工艺下的 APD 像素结构。该 APD 在深 n 阱中实现,多个区域被一圈 p^+ 保护环所包围。在 Geiger 模式中,APD(本书中光电二极管被称作单光子雪崩二极管,SPAD)产生一个峰状信号,而不是一个模拟输出。如图 4.6 所示,一个反相器作为脉冲整形电路,用来将输出的峰状信号转化为数字脉冲。入射的光信号与产生的脉冲数成正比。

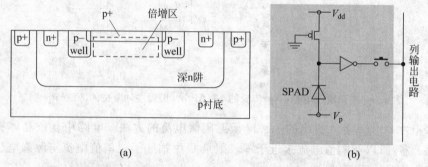

图 4.6 (a)使用标准 CMOS 工艺的基本 APD 结构;(b)像素中 Geiger 模式的 APD 的电路。PMOS 连接到 V_{dd} 上作为淬火过程中的电阻;V_p 是一个负电位,使 PD 进入雪崩区

4.3 高速度

尽管 CMOS 图像传感器的每列电路的处理时间相对较慢,约 $1\mu s$,但是 CMOS 图像传感器具有列并行性,列并行电路能够实现高的数据速率,所以高速是其一项优势。以前,超高速摄像机通常基于 CCD 技术[289],然而近期几种基于 CMOS 图像传感器的高速摄影机已研发出来。高帧频速率可以弥补 CMOS 图像传感器中滚筒快门的固有缺陷。

4.3.1 全局快门

为了获得速度大于 1000fps 的超高速图像,采用全局快门是非常有必要的。通常情况下,将一个晶体管和一个电容添加到 3T-APS 像素中来实现全局快门功能[291,298,299]。像素电路如图 4.7 所示,图像拖尾对于超高速图像是非常严重的问题,因为 4T-APS 需要相对较长的传输时间,所以它不适于用来实现全局快门功能,另外,在 3T-APS 和 5T-APS 中,为了确保没有图像拖尾,硬复位是必要的[146]。

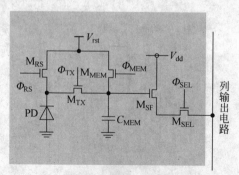

图 4.7 全局快门功能的基本像素电路[291]

4.4 宽动态范围

4.4.1 宽动态范围的原理

人眼具有约200dB的宽动态范围。为了实现这样的宽动态范围,人眼有三种机制[300]。其一,人的眼睛具有两类感光细胞:视锥细胞和视杆细胞,它们对应于两类具有不同光敏度的光电二极管。其二,眼睛的感光响应曲线呈对数形式,致使达到饱和的速度慢。其三,响应曲线根据环境光级别或平均亮度级别发生变化。相比之下,常规的图像传感器只有60~70dB的动态范围,这主要受光电二极管的阱容量限制。

在汽车和安防等应用中,要求图像传感器的动态范围超过100dB[275]。为了扩大动态范围,许多方法已经被提出并论证,它们可以被分为三类:非线性响应、多次采样和饱和检测。图4.8对这些方法进行了说明。图4.9显示了一张通过宽动态范围的图像传感器拍摄的示例图,该图像传感器是由静冈大学的S. Kawahito和他的同事们研发的[301]。左侧极亮的灯泡以及右侧黑暗条件下的物体均可被看到。

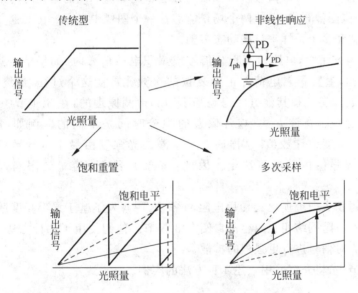

图4.8 提高动态范围的基本概念

如上所述,人的视网膜具有高灵敏度和低灵敏度的两种类型感光体。带有高灵敏度PD和低灵敏度PD的图像传感器可以实现很大的动态范围。这种图像传感器已经用CCD技术实现[336]。在文献[302]报道的CMOS图像传感器中,有两种高、低灵敏度的光电探测器,其中一个FD作为低灵敏度的光电探测器。虽然FD处于遮光状态,但仍然可以收集高光照条件下光生电荷,根据这种原理,该传感器在标准图像格式下可实现110dB的内部场景动态范围。

非线性响应是一种将光响应从线性修改到非线性的方法,例如对数响应。该方法被分为两种,分别为采用对数传感器和阱容量调整。在对数传感器中,光电二极管具有对数响应。通过调整阱容量,可以实现非线性响应,但在某些情况下,也可以实现线性响应,这些将

图 4.9 示例图是由 S. Kawahito 等人开发的宽动态范围图像传感器所拍摄的[301]。该图像由多个图像合成产生,在 4.4.4 节中进行详细介绍(在此向静冈大学的 Kawahito 教授致谢)

在后面进行介绍。

多次采样是一种信号电荷被多次读取的方法。例如以不同的曝光时间来获取亮或暗的图像,然后将两张图像合成,以使两个场景能够在一个图像中显示。在文献[303]中,详细分析了通过阱调整和多个采样来扩展动态范围。

在饱和探测方法中,电路检测积分信号或累积电荷信号时,如果信号达到阈值,那么电路进行的操作,重置积累电荷并记录复位的次数。重复这个过程,最终得到残余电荷信号和复位次数。有几种其他方法也是基于饱和检测原理的,在 3.4 节中讨论过的脉冲调制就是其中之一。在该方法中,各像素的积分时间是不同的。例如,脉冲宽度调制中,输出为脉冲宽度或计数值,这样最大可检测的光强度由最小可计数值或时钟决定,同时最小检测光强值由暗电流决定。因此,该方法不受阱容量的限制,具有很宽的动态范围。

德国佛罗恩霍夫研究所已经使用电阻网络[304,305]开发了基于局部亮度适应的宽动态范围图像传感器。电阻网络相关内容请参阅 3.3.3 节。他们使用了一种方法,通过该网络可以在明亮的地方对信号进行更强烈的扩散。

在以下章节中,将介绍采用上述 4 种方法的实例。

4.4.2 双重敏感性

当两类具有不同灵敏度的光检测器被集成在一个像素时,传感器可以覆盖大范围的照明强度。在明亮的环境下,使用具有较低灵敏度的 PD,而在黑暗的条件下,使用具有较高灵敏度的 PD。这类似于人类的视觉系统,如上所述,并且已经通过 CCD[336]实现。在文献[302]中,FD 被用作低灵敏度光探测器。在明亮的光照条件下,衬底上产生一些载流子并扩散到 FD 区,成为最终信号的一部分。这个结构是在文献[306]中首次提出的。这是一种直接探测光强的方法,所以不会出现多次采样的图像采样延迟。另一种实现双光电探测器的方法在文献[307]中提到,该方法采用一个 PG 作为主要的 PD,采用 n 型扩散层作为第二个 PD。

4.4.3 非线性响应

4.4.3.1 对数传感器

现在将对数传感器用于宽动态范围成像,可以获得超过100dB的动态范围[163-169]。

对数传感器的问题包括制造过程的参数变化,在低光强度下的较大噪声,以及图像拖尾。这些缺点主要是由于器件在亚阈值工作状态下以扩散电流为主。

已经有一些关于传感器中的对数实现和线性响应的报道[308,310]。线性响应在暗光区更好,而对数响应则更适合在明亮的光照区域。在这些传感器中,过渡区中的矫正是必不可少的。

表4.1 智能CMOS图像传感器的动态范围

分类	实现方法	参考文献
双灵敏度	一个像素中含PG和PD	[302,306,307]
非线性响应	对数传感器	[163-169]
	对数/线性响应	[308-310]
	阱容量调整	[298,311-318]
	控制集成时间	[319,320]
多重抽样	双重采样	[321-323]
	带有固定的短曝光时间的多重抽样	[324,325]
	带有变化的短曝光时间的多重抽样	[301]
	带有像素级ADC的多重抽样	[140,184,186]
饱和检测	本地集成时间和增益	[326]
	饱和计数	[263,326-332]
	脉冲宽度调制	[179,263,333]
	脉冲频率调制	[141,194,196,202,334,335]
亮度扩散	电阻网络	[304,305]

4.4.3.2 阱容量调整

阱容量调整是一种在曝光过程中控制电荷收集区的势阱深度的方法。在该方法中,用一个晶体管漏极来接受溢出电荷。通过控制位于电荷收集区和溢出漏极之间的栅极,可以使光响应曲线变为非线性。

图4.10给出的是采用溢出漏极来提高最大阱容量的像素结构。当强光照射到传感器时,光生载流子在PD阱达到饱和且涌入FD节点。通过逐渐减小溢出漏端(OFD)的势阱,强光信号不会使收集区达到饱和,弱光信号也可被检测到。

这种方法实现了与3T和4T有源像素传感器几乎相同的像素结构的饱和响应,因此,在弱光照条件下它具有良好的信噪比。这种方法的缺点是溢出机制消耗了像素面积,所以填充因子减少了。

该方法可以在3T有源像素传感器[311,312,319]以及4T有源像素传感器[298,313-316]中实现。因为4T-APS的灵敏度比3T-APS的好,因此从暗光到亮光的动态范围就可以得到改善。

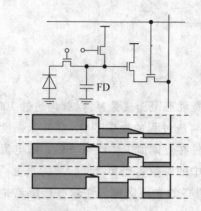

图 4.10 带有溢出漏极的宽动态范围图像传感器的像素结构[311]

在文献[314]中提出了一种通过 CDS 来降低噪声的方法,其随机噪声和列固定模式噪声都仅为 0.15mVrms,实现了相当高的信噪比,该研究采用了一个横向 OFD(溢出漏极)的层叠电容。文献[315]结合文献[314]提出的结构,通过引入光电流直接输出模式,实现了超过 200dB 的超高动态范围。文章在光电流的直接输出模式中,采用了对数响应模式,进一步提高了动态范围。

北海道大学的 M. Ikebe 在文献[317]中,提出并验证了通过负反馈重置来实现 PD 电容调制的方法。此方法不改变像素结构,采用传统的 3T 或 4T APS,通过一列差分放大器来控制复位电压,这种方法不仅可以实现宽动态范围,还能够抑制噪声。

在文献[318]中,通过在一个像素中结合 3T-APS 和一个 PPS 来提高动态范围。一般来说,与 PPS 相比,APS 在低光照下具有更优越的信噪比(SNR),也就是说,PPS 在高光强范围的性能还是可以接受的。而 PPS 适合与 OFD 共同使用,因为一个列电荷放大器在 PD 和 OFD 中都可以完全传输电荷。在这种情况下,就不需要关心 PD 信号的电荷是否被转移到 OFD 上了。

4.4.4 多次采样

多次采样是指对信号电荷进行多次读取,并将读取出的多个图像合成一个图像。通过这种方法,可以很方便地实现宽动态范围。然而,在合成所获得图像时存在着一些问题。

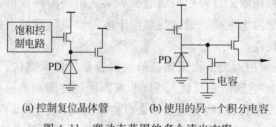

(a) 控制复位晶体管　　(b) 使用的另一个积分电容

图 4.11 宽动态范围的多个读出方案

4.4.4.1 双采样

在双采样技术中,一个芯片中采用了两种读出电路,如文献[321,323]。对第 n 行中的像素数据进行采样,并保存在一个读出电路中,然后复位像素。而此时,采样第 $(n-\Delta)$ 行中的像素数据,并将其保存在另一个读出电路,然后复位该像素。阵列的大小为 $N \times M$,其中 N 是行数,M 是列数。在这种情况下,首先读出的行的积分时间为 $T_1 = (N - \Delta) T_{\text{row}}$,然后读出的行的积分时间为 $T_s = \Delta \times T_{\text{row}}$。$T_{\text{row}}$ 为读出一行数据所需要的时间,$T_{\text{row}} = T_{\text{SH}} + MT_{\text{scan}}$,$T_{\text{SH}}$ 为采样保持(S&H)时间,T_{scan} 为从采样保持电路读出所需的时间(或者说扫描数据的时间)。这里定义一帧的时间 $T_f = NT_{\text{row}}$,T_1 可以表示为

$$T_1 = T_f - T_s \tag{4.1}$$

因此,动态范围为

$$\text{DR} = 20\log \frac{Q_{\max} T_1}{Q_{\min} T_s} = \text{DR}_{\text{org}} + 20\log\left(\frac{T_f}{T_s} - 1\right) \tag{4.2}$$

式中,DR_{org} 为未采用双采样的动态范围,而 $Q_{\max}$ 和 $Q_{\min}$ 分别是最大和最小的积分电荷。例如,如果 $N = 480$,$\Delta = 2$,则积分比例 $T_f/T_s \approx T_1/T_s$,约为 240。因此,它大约可以扩大 47dB 的动态范围。

这种方法只需要两个 S&H 区,而对像素结构不做任何改变,因此,它可以应用于具有高灵敏度的 4T-APS。而这种方法的缺点是只有两个积分次数,在两次不同曝光之间的边界会出现比较大的信噪比落差。

在两个不同曝光的边界时,积累的信号电荷从它的最大值 $Q_{\max}$ 变为 $Q_{\max} T_s/T_1$,这会导致较大的 SNR 落差。信噪比落差 ΔSNR 为

$$\Delta\text{SNR} = 10\log \frac{T_s}{T_1} \tag{4.3}$$

如果在这两个区域中,噪声水平没有改变的话,那么上述例子中的 ΔSNR 约为 -24dB ($T_s/T_1 \approx 240$)。

4.4.4.2 多次采样

固定的短时间曝光。 为了减小信噪比落差,M. Sasaki 等人在文献[324,325]中提出采用多个短时间的曝光方法。多次采样方法必须以像素非破坏性读出为前提。将短的积分时间 T_s 曝光得到的信号读取 k 次,就可以使信噪比落差变为

$$\Delta\text{SNR} = 10\log k \frac{T_s}{T_1} \tag{4.4}$$

同时,T_1 更改为

$$T_1 = T_f - kT_s \tag{4.5}$$

因此,动态范围的提升值 ΔDR 为

$$\Delta\text{DR} = 20\log\left(\frac{T_f}{T_s} - k\right) \tag{4.6}$$

如果 $T_f/T_s \approx 240$,而 $k = 8$,则 ΔDR ≈ 47dB,ΔSNR $= -15$dB。

渐变的短时间曝光。 在之前的方法中,多次读取的短曝光的时间是固定的。在文献[301]中,M. Mase 等人通过改变短曝光的时间改进了该方法。在较短的曝光时间段,采用几个不同的短曝光时间。在短曝光时间中插入一个更短的曝光时间,同时在这个更短的曝光时间内再插入一个进一步缩短的曝光时间,以此类推,如图 4.12 所示。该方法可通过具有列并行 cyclic ADC 的快速读出电路实现。在该方法中,动态范围扩展 ΔDR:

$$\Delta \text{DR} = 20\log \frac{T_\text{f}}{T_\text{s,min}} \tag{4.7}$$

其中，$T_\text{s,min}$ 为最短的曝光时间。采用这种方法，通过使每一次曝光时间中的 T_s/T_l 最小来降低 SNR 的落差。

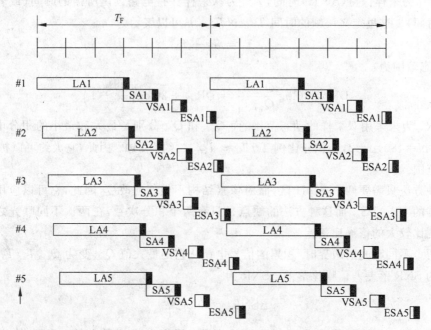

图 4.12 渐变的短曝光时间的曝光过程和读出时序[301]。LA：长时间积分；SA：短时间积分；VSA：很短的时间积分；ESA：极短的时间积分

像素级 ADC。另一种实现多次采样的方法是通过像素级 ADC，如文献[140,184,186]所述。在这种方法中，每 4 个像素中采用一个具有单斜坡的位串行的 ADC。ADC 的精度随着积分时间的变化而变化，从而获得高的分辨率。对于较短的积分时间，ADC 采用较高的精度。文献[186]中提出的图像传感器集成了一个 DRAM 的帧缓冲器，在像素数为 742×554 的基础上，实现了超过 100dB 的动态范围。

4.4.5 饱和探测

饱和探测是一种基于监测和控制饱和信号的方法。由于该方法是异步的，使得每个像素很容易实现自动曝光，常见的一个问题是如何降低复位噪声。由于该方法中有多个复位操作，所以很难使用噪声消除机制。另外，复位时信噪比的降低也是一个问题。

4.4.5.1 饱和计数

当信号电平达到饱和时，电荷收集区域被复位，并且重新开始积分。通过重复该过程，并计算一段时间内的复位计数，该时间段的总电荷可以通过复位次数以及最后残余电荷信号来计算[263,326-332]。文献[263,327-329]在像素级实现计数电路，而文献[326,330,331]则在列级实现。在像素级实现的方法中，计数电路需要消耗额外的面积，这会导致填充因子降低。文献[263,327]中提出的 TFT 技术可以缓解这个问题，但是需要采用特殊的工艺。而无论是像素级还是列级,采用饱和计数这种方法都需要帧存储器。

4.4.5.2 脉冲宽度调制

通过脉冲宽度调制来提高动态范围的方法已经在 3.4.1 节中提到,文献[179]采用这种方法实现了图像传感器动态范围的提高。

4.4.5.3 脉冲频率调制

通过脉冲频率调制来提高动态范围的方法已经在 3.4.2 节中提到,文献[141,192-194]采用这种方法实现了图像传感器动态范围的提高。

4.4.6 亮度扩散

在这种方法中,输入光由电阻网络扩散,正如 3.3.3 节以及文献[304,305]所提到的。亮点的输出通过扩散被抑制,从而使光响应曲线变为非线性。由于电阻网络的响应速度不是很快,因此这个方法难以用于拍摄快速移动的物体。

4.5 解调

4.5.1 解调的原理

在解调方式中,调制光信号照射在目标物体上,反射的光被图像传感器捕获。这种工作方式下,传感器只侦测调制信号,去除了所有的静态背景噪声,从而可以有效地获得高信噪比的信号。实现该技术的图像传感器需要一个调制光源,这种传感器可以在几乎不受背景环境光线影响的条件下得到满意的图像,这使得它在智能交通系统、工厂自动化和机器人技术领域具有很好的应用前景。除了上述应用,传感器还可以用于跟踪由调制光源指定的一个目标,举例来说,这种传感器可以很容易实现不同光照条件下的动作捕捉。该解调技术的另一个重要应用是基于时差测距方法实现的三维测距仪,相关具体细节在 4.6 节中详述。

解调技术在传统的图像传感器中是很难实现的,因为传统的图像传感器工作在积累电荷的模式下,所以调制信号很容易被积累调制电荷弱化。基于解调技术的智能图像传感器的概念在图 4.13 中加以说明。照明光 $I_0(t)$ 由频率 f 调制,反射(或散射)光 $I_r(t)$ 也由频率 f 调制。传感器接收到的光线由反射光 $I_r(t)$ 与背景光 I_b 组成,光电二极管的输出是与 $I_r(t)+I_b$ 成正比例的。光电二极管的输出要在上式的基础上乘以同步调制信号 $m(t)$ 并进行积分得到。输出电压为[337]

$$V_{\text{out}} = \int_{t-T}^{t} (I_r(\tau) + I_b) m(\tau) d\tau \qquad (4.8)$$

其中 T 是积分时间。

基于图 4.13 的概念,已经有数项研究在智能图像传感器中实现了解调功能。实现方式可以分为两类:相关法(correlation)[166,337-344]和在一个像素中采用双电荷存储区的方法[345-353]。图 4.13 中所说明概念最直观简单的实现方式是相关法。

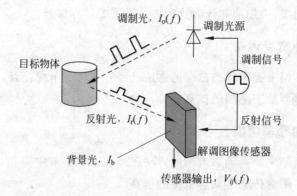

图 4.13 解调图像传感器的概念

4.5.2 相关法

相关法的基本原理是将侦测信号与参考信号相乘并进行积分，或进行低通滤波。相关法的工作过程由式(4.8)描述。图 4.14 展示了相关法的概念。实现相关法最关键的要素是乘法器。在文献[344]中使用了简单的源连接型乘法器[42]。文献[166]中运用了吉尔伯特单元(Gilbert Cell)消除背景光信号。

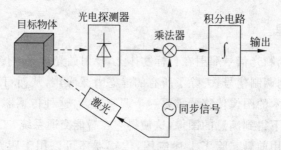

图 4.14 相关法的概念示意图

在相关法的应用中，更适合使用三相位参考以便通过振幅与相位来获得足够的调制信息。图 4.15 展示了可以实现三相位参考的像素电路图。具有三个输入的源连接电路可以实现此目的[344]。此结构可以完成调幅与调频的解调。

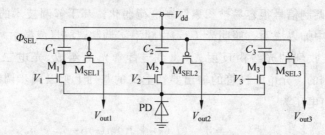

图 4.15 相关法中所采用的像素电路(以文献[344]为例)

电路中各个 M_i 的漏极电流 I_i 可表示为

$$I_i = I_0 \left(\frac{e}{mk_BT} V_i \right) \tag{4.9}$$

式中所用符号与附录 E 中所使用的相同。那么每个 M_i 的漏极电流又可表示为

$$I_i = I \frac{\exp\left(\frac{eV_i}{mk_BT}\right)}{\exp\left(-\frac{eV_1}{mk_BT}\right)+\exp\left(-\frac{eV_2}{mk_BT}\right)+\exp\left(-\frac{eV_3}{mk_BT}\right)} \quad (4.10)$$

由此可以得到相关等式

$$I_i - \frac{I}{3} = -\frac{e}{3mk_BT}I(V_i - \bar{V}) \quad (4.11)$$

其中 $\bar{V}$ 是 V_i 的平均值($i=1,2,3$)。

此方法已被应用到三维测距仪[340,350]和频谱匹配装置中[354]。

4.5.3 双电荷存储区法

在一个像素中使用双电荷存储区的方法与相关法的原理本质上是一样的,但是更易于实现,具体原理如图 4.16 所示。此方式中的调制信号为脉冲信号或开关信号。当调制信号为开时,信号存储到其中的一个存储区,相关法是由此操作过程完成的。当调制信号为关时,信号存储到另外一个存储区,从而将两个信号相减以消除背景信号。

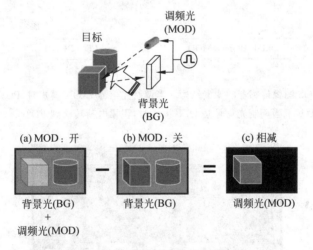

图 4.16 双存储区法的概念示意图

图 4.17 展示了一个具有双存储区的像素结构[348]。图 4.18 展示了像素的版图,使用 3 层金属和 2 层多晶硅 $0.6\mu m$ 标准 CMOS 工艺。像素结构类似于包含两个读出电路的光栅型有源像素(PG APS),其中含有两个传输栅(TX1 和 TX2)和两浮空节点(FD1 和 FD2)。光栅结构代替光电二极管作为光电探测器,分别通过 TX1、TX2 与 FD1、FD2 相连。两个读出电路共用一个复位晶体管(RST)。两个输出信号 OUT1 与 OUT2 相减,最后只得到一个调制的信号。有的研究者还提出了另外一种具有两个存储区的相似结构[345,346,353]。

图 4.19 给出了传感器的工作时序图。首先,当调制光信号关闭时,通过使 RST 导通完成复位操作。当调制光信号打开时,对光栅进行偏置,进行光载流子的积累。然后调制光信号关闭,光栅关闭,通过打开 TX1 将积累的电荷转移至 FD1。以上过程是调制信号为打开时的操作过程,在这个阶段调制光信号和静态光信号都被储存在 FD1 中。接下来,调制信号关闭,光栅再次被偏置,开始收集光载流子。在调制信号的关闭状态结束时,将积累电荷

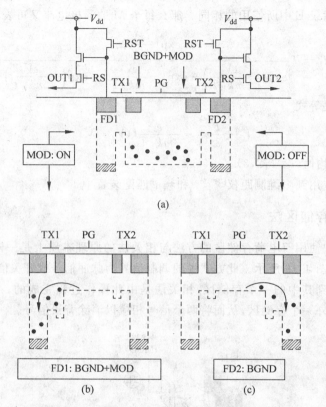

图 4.17 解调 CMOS 图像传感器的像素结构。当调制光信号为 ON 模式时,PG(a)中的积累电荷转移到 FD1(b);当调制光信号是 OFF 模式时,积累电荷转移到 FD2(c)

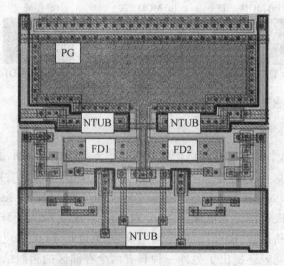

图 4.18 解调图像传感器像素的版图(像素尺寸为 $42\mu m \times 42\mu m$)

转移至 FD2,只有静态光信号被储存在 FD2 中。重复上述过程,则调制信号开启状态与关闭状态积累的电荷分别储存在 FD1 与 FD2 中。根据积累的电荷量,FD1 与 FD2 的电压会呈现阶梯状的下降。通过在一个规定的时刻测量 FD1 与 FD2 的压降的差值就可以提取出调制信号。

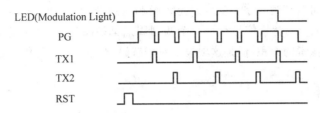

图 4.19 解调图相传感器的时序图[348]。PG：光栅；TX：传输栅；RST：复位管

图 4.20 所示的是使用该图像传感器的实验结果[343]。调制光只照射两个物体（猫和狗）其中的一个，解调图像只显示被调制光照射的猫。

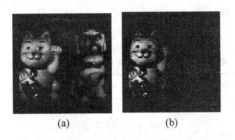

图 4.20 （a）普通图像；（b）解码图像

图 4.21 展示的是进一步的实验结果。在实验中，一个调制的 LED 光源固定在一只四处移动的狗的脖子上，解调图像只显示了该调制 LED 光源，进而给出了物体的移动轨迹，这说明这个解调图像传感器可以用于目标追踪。另一个应用是在相机系统中抑制饱和[355]。

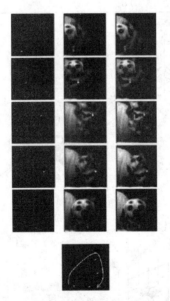

图 4.21 标记追踪的实例。图像按时间顺序由顶到底排列。左列：传感器提取的调制光图案；右列：背景光与调制光一起输出；中间列：只含有背景光的输出；最底部的图案显示了物体的运动轨迹。为方便起见，将运动轨迹与 LED 光源轨迹重叠[343]

这种工作方式可以获得几乎不受背景光线影响的图像,然而动态范围仍受存储区电容的制约。文献[350,351]中,背景信号在每一个调制周期都被减掉,实现了宽动态范围。尽管增加的电路需要消耗像素的面积,但是这项技术对于解码型 CMOS 图像传感器还是很有用的。

4.6 三维测距

工厂自动化(FA)、智能交通系统(ITS)、机器人视觉、手势识别等都是测距很重要的应用方向。通过利用智能 CMOS 图像传感器可实现三维(3D)测距仪或者与距离有关的图像捕捉。研究者研究了几种适用于 CMOS 图像传感器的测距方法,它的原理是基于时差测距(TOF)、三角测距等方法,总结在表 4.2 中。图 4.22 显示了一张 3D 测距仪拍摄得到的、直接显示物体距离的图像[356]。

表 4.2 三维测距的智能 CMOS 图像传感器

分类	实现方法	从属和参考资料
直接 TOF	APD 阵列	ETH[97,98],MIT[357]
间接 TOF	脉冲	ITC-irst[358,359],Fraunhofer[360,361],
	正弦曲线	SizuokaU.[362],PMD[340,341,363,364],CSEM[353,365],Canesta[366],JPL[367]
三角法	双目	Shizuoka U.[368],Tokyo Sci. U.[369],Johns Hopkins U.[370,371]
	结构光	Camegie Mellon U.[87,372],U. Tokyo[350,356,373,374],SONY[157,375],Osaka EC. U.[342]
其他	光强度(光强深度)	Toshiba[376]

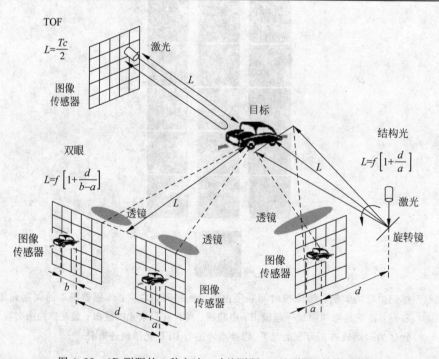

图 4.22 3D 测距的 3 种方法:时差测距,双目测距,光截面测距

4.6.1 时差测距

时差测距(TOF)是通过测量光信号往返传感器与被测物体的时间来测距的方法,在测光和测距(LIDAR)领域已经应用了很多年[377]。传感器与物体之间的距离 L 可表示为

$$L = \frac{Tc}{2} \tag{4.12}$$

T 是往返时间(TOF=$T/2$),c 是光速。时差测距的最显著特点是它的系统很简单,只需要一个时差测距传感器和一个光源。时差测距传感器分为直接时差测距与间接时差测距。

4.6.1.1 直接时差测距

直接测距是传感器内部的每一个像素都直接测量往返时间,相应地,它需要一个高速的光电探测器和高精度的计时电路。例如,对于 $L=3\mathrm{m}$,$T=10\mathrm{ps}$。为了获得毫米级的精度,必须要进行取平均操作。直接时差测距的优势是它的测距范围很宽,可从数米到数千米。

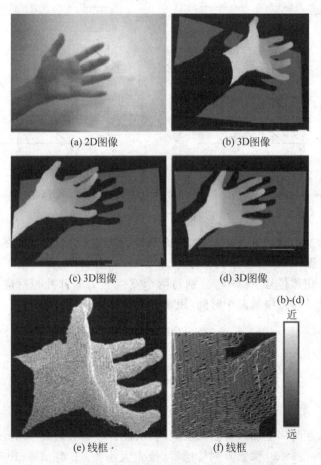

图 4.23 由 Y.Oike 研发的 3D 测距仪所拍摄的照片(在此向 Y.Oike 博士致谢)

直接时差测距传感器使用的高速光电探测器是工作在盖革模式(Geiger Mode)下、基于标准 CMOS 工艺技术的有源像素[97,98,357],详细内容见 2.3.4 节。研究人员已经采用带高压工艺的 0.8μm 标准 CMOS 工艺技术制成拥有 32×32 个像素、每个像素集成了图 4.6 中的电路、每个像素面积为 58μm×58μm 的传感器。有源像素的正极偏置在 −25.5V 的高压。像素的跳动时间为 115ps,如果要获得毫米级精度,必须进行平均操作。在距离大约为 3m、10^4 个测量深度的精度下可以实现仅为 1.8mm 的测量距离偏差。

4.6.1.2 间接时差测距

为了降低直接时差测距对传感器的要求,发展了间接时差测距传感器[340,341,345,353,358-361,363-367]。在间接测距传感器中,往返时间不是直接被测量的,而是使用两个调制光源信号来进行间接测量。间接时差测距传感器中每个像素都有两个电荷积累区域以完成信号的解调,具体内容见 4.5 节。图 4.24 给出了间接时差测距传感器的时序图,图中展示了两个例子。

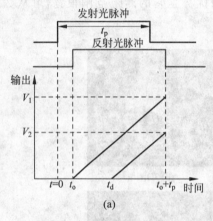
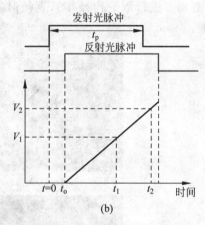

图 4.24 工作于两种不同脉冲的间接时差测距的时间示意图:(a)第二个脉冲与第一个脉冲之间有 t_d 的时间延迟;(b)两个持续时间不同的脉冲

调制信号是脉冲信号或开关信号时,两个脉冲以既定的 MHz 级别的频率发射,两个脉冲之间延迟为 t_d。图 4.25 给出了其工作原理[358,359],在这种工作方式下,时差信号通过以下方法获得。两个积累信号 V_1 与 V_2 分别与两个宽度为 t_d 的脉冲相对应,如图 4.24 所示。根据延迟时间 t_d 以及时差中的两个时间,距离 L 可以表示为

$$t_d - 2 \times \text{TOF} = t_p \frac{V_1 - V_2}{V_1} \tag{4.13}$$

$$\text{TOF} = \frac{L}{c} \tag{4.14}$$

$$L = \frac{c}{2}\left[t_p\left(\frac{V_2}{V_1} - 1\right) + t_d\right] \tag{4.15}$$

文献[359]中,在时序上加入背景减法周期,从而可以滤去最高 40klux 的背景光。文献[359]中研发了一种 50×30 像素阵列传感器,像素大小为 81.9μm×81.7μm,使用 0.35μm 标准 CMOS 工艺技术制造,可以在 2~8m 的范围内,实现测距 4% 的精度。

另一种使用两个不同持续时间的脉冲来测量时差的间接时差测距传感器[360-362]。估算

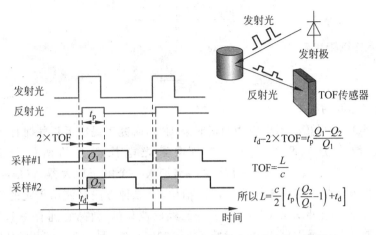

图 4.25 使用两个延迟脉冲进行间接时差测距的工作原理

L 的过程如下所示。V_1 与 V_2 是曝光时间分别为 t_1 与 t_2 时的输出信号电压。如图 4.24(b)所示,根据这 4 个参数,可以是实现相交点 $t_0 =$ TOF 的差值。因此

$$L = \frac{1}{2}\left(\frac{V_2 t_1 - V_1 t_2}{V_2 - V_1}\right) \tag{4.16}$$

接下来,可以在间接时差测距中用正弦波信号代替脉冲信号[340,341,353,363-367]。

如图 4.26 所示,时差测距可以通过对 4 个点进行采样,采样完成后计算相移 φ 的方式来估算距离,其中两个采样点之间相差 $\pi/2$[363]。4 个采样值 $A_1 \sim A_4$ 由相位移 φ、幅值 a 与固定漂移量 b 表示为

$$A_1 = a\sin\phi + b \tag{4.17}$$

$$A_2 = a\sin(\phi + \pi/2) + b \tag{4.18}$$

$$A_3 = a\sin(\phi + \pi) + b \tag{4.19}$$

$$A_3 = a\sin(\phi + 3\pi/2) + b \tag{4.20}$$

通过求解上述方程组,可解得 φ、a、b 为

$$\phi = \arctan\left(\frac{A_1 - A_3}{A_2 - A_4}\right) \tag{4.21}$$

$$a = \frac{\sqrt{(A_1 - A_3)^2 + (A_2 - A_4)^2}}{2} \tag{4.22}$$

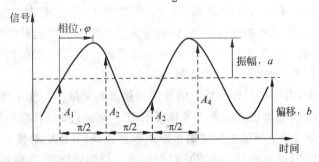

图 4.26 使用正弦发射光新信号进行间接时差测距的工作原理[363]

$$b = \frac{A_1 + A_2 + A_3 + A_4}{4} \tag{4.23}$$

最后,通过 φ 来计算距离 L 为

$$L = \frac{c\phi}{4\pi f_{\text{mod}}} \tag{4.24}$$

其中 f_{mod} 为调制光源的频率。

为了在 CMOS 图像传感器中应用间接时差测距,研究者提出了几种拥有双电荷储存区的像素。这些像素分为两类:一种是在像素光电探测器的两侧分别放置一个 FD[340,341,345,353,362,363,365,366];另一种是在像素内部加入一个电压放大器,将信号储存在电容上。总之第一种方法是改变光电探测的器件结构,第二种方法是使用传统的 CMOS 电路。光混频器(photomixing device,PMD)是一种已经商业化的器件,它拥有一个 PG,并且 PG 两侧有两个存储区[363],最大可用的像素数目为 160×120。另一种商业化的间接时差测距传感器拥有 64×64 的像素阵列和高速的时钟发生器和 ADC[366]。文献[366]与文献[362]中的传感器由标准 CMOS 工艺(或在其基础上进行微小修改)制成。值得特别说明的是,文献[362]中的传感器拥有 QVGA 阵列,在智能交通系统及手势识别领域有很广泛的应用。

4.6.2 三角测距

三角测距是一种使用三角形的几何排列测量到视场(FOV)距离的方法。三角测距的方法主要分为两类:主动式和被动式。被动式也称为双目法或立体视觉方法,在该方法中有两个传感器。主动式称为结构光源方法,在该方法中使用图案化的光源来照亮视场。

4.6.2.1 双目法

被动式的优势在于它不需要外加光源,只需要两个传感器就能实现测距。数项相关传感器的研究已经被发表[368-371]。两个传感器分别集成到两套成像区域来实现立体视觉,这意味着两个传感器必须能够对同一视场进行识别。对于一个典型的场景来说,两个传感器中的视场很复杂。文献[368]中的传感器在工作时已经规定了一个已知物体的视场,三维信息可以提高物体的识别率。这种传感器拥有两个成像区和一套包含 ADC 的读出电路。文献[371]中的传感器集成了电流模式差动运算电路。

4.6.2.2 结构光源法

现在已经有很多关于主动法的研究[87,157,342,350,356,372-374]。主动式工作时通常需要一个结构光源,结构光源需要对视场进行扫描,它通常是条状的。为了能够将需要的光图案从周围光线中识别出来,我们需要大功率的光源及扫描系统。文献[372]提出了一种基于 $2\mu m$ CMOS 工艺的集成传感器系统,其像素阵列为 5×5。

在结构光源法中,需要寻找拥有最大信号值的像素。文献[87]使用赢家通吃(WTA)电路来寻找最大值,WTA 电路已在 3.3.1 节做了详述。文献[356,373]中传统的 3T 有源像素可以在两种模式下工作:普通的图像输出模式和脉宽调制(PWM)模式,如图 4.27 所示。PWM 已在 3.4 节做了详述,在这里 PWM 可以用作 1 位的 ADC。在传感器中,每一列放置一个 PWM-ADC,这样就实现了一个列并行 ADC。图 4.27(b)中,像素就像传统的 3T 有源

像素那样工作,但是在图 4.27(c)中输出总线是预充电的,像素的输出信号通过列放大器与一个参考电压 V_{ref} 进行比较。像素的输出电压与输入的光强成正比,这样就完成了 PWM。使用 3T 像素结构,传感器可以达到 VGA 级别的大阵列,实现了高精度,在所测距离为 1.2m 时,精度可达到 0.26mm。文献[157]与文献[375]的研究通过设计模拟电流复制器[378]和比较器能够快速地判定最大峰值。文献[157]中将模拟操作电路集成到每个像素中,但是在文献[375]中只集成了 4 组模拟帧存储器来减小像素的尺寸,并且使用 4T 像素结构实现了彩色的 QVGA 阵列。

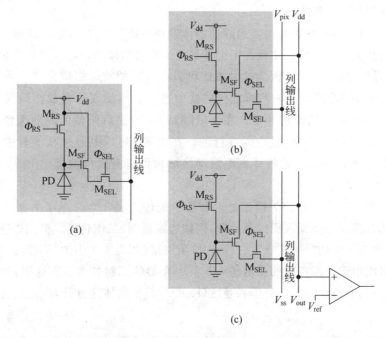

图 4.27 (a)传统 3T 像素结;(b)修改后工作在普通成像模式;(c)脉宽调制工作模式,详见文献[373]

在结构光源法中也需要抑制周围光源的影响使结构光源照射到视场的光更容易被辨别出来。可以使用对数传感器与调制技术以实现更大的动态范围。文献[350]中研制的传感器,其动态范围为 96dB,同时其信号-背景比(SBR)达到了 36dB。

4.6.3 关键值深度法

关键值深度指的是焦深、灰度等[379]。这种方法用来测量一些深度物理量的距离。根据文献[380]的论述,这种方式经常用于相机系统中。对于几张具有不同焦距的图片,一个焦深传感器可以计算出所有像素的下列拉普拉斯基础值 d。

$$d = |\, l_F(x+1,y) + l_F(x-1,y) - 2l_F(x,y)\,| \\ + |\, l_F(x,y+1) + l_F(x,y-1) - 2l_F(x,y)\,| \tag{4.25}$$

根据在 $0.6\mu m$ 标准 CMOS 工艺下的仿真结果,当拍摄速度达到 1500fps 时,系统可以以 30fps 的速度输出深度图像(64×64 像素,50 级深度阶度)[380]。

关于灰度方法,也有研究成果发表[376]。一个灰度深度传感器通过判断物体反射光线的光强来估算距离,所以估算的距离会受到物体反射系数的影响。这种方法很容易实现。

4.7 目标追踪

目标追踪是一种可以对特定的物体进行追踪的技术。对于智能 CMOS 图像传感器来说,目标追踪是一个很重要的应用,它需要进行实时的信号处理。举例来说,在机器人视觉应用中,需要快速、紧凑的信号处理并且要保证低功耗。目前有很多这种类型的智能 CMOS 图像传感器研发出来。本节中,我们将能够实现目标追踪的图像传感器分成几类,并分别介绍它们的工作原理。

为了实现目标追踪,必须先将要追踪的物体识别并提取出来。因此,找到追踪物体的矩心很重要。有相关的研究可以实现在一个场景中识别出物体的矩心。在获得物体的矩心的方法中,帧差操作是很有效的,因为这种方法可以把移动的物体提取出来。为了实现帧差操作,需要有帧的缓冲器。在一些智能传感器中通常将帧的缓存器放在像素内部[328,381],这样外部就不需要额外的帧缓冲器。在文献[328]中通过对两帧之间的像素值做差值来寻找运动物体,差值操作是在芯片内进行的。在文献[382]中通过在像素中加入一个电容和一个比较器来检测运动的物体。电阻网络结构可以有效检测运动[42,45],在 3.3.3 节中已详述。

目标跟踪的传感器主要分为模拟处理和数字处理。表 4.3 总结了一些方法的典型例子。每种方法都需要一些输入图像预处理,例如边缘检测、二值化等。模拟处理可以使用最大值检测(如表 4.3 中的"MaxDet")、投影和电阻网络(如表 4.3 中的"RN")。

另一种可以用于目标追踪的技术是 4.5 节中介绍的调制技术。在这里,虽然需要一个调制光源,但是很容易在场景中将感兴趣区(ROI)与其他物体区分开,因此实现目标追踪相对容易。

表 4.3 目标追踪的智能 CMOS 图像传感器

方法	预处理	工艺/μm	像素♯	像素尺寸/μm	速度	功耗	参考资料
最大值检测法	WTA 和 ID-RN	2	24×24	62×62	7kpix/s		[383]
最大值检测法	ID-RN 和比较器	0.6	11×11	190×210	50μs(sim.)		[384]
投影法	电流和	2	256×256	35×26	10ms		[173]
投影法	边缘检测	2	凹陷 9×9,圆周 19×17	凹陷 150×150,圆周 300×300	101pix/s	15mW	[385]
投影法	二进制化	0.18	80×80	12×12	1000fps	30mW	[386]
投影法	充电求和	0.6	512×512	20×20	1620fps	75mW	[387]
二维电阻网络		0.25	100×100 (hex.)	87×75.34	~7μs	36.3mW	[45]
胞状类神经网络		0.18	48×48	85×85	500μs,~725MIPS/mw	243μW	[388]
数字处理		0.5	64×64	80×80	1kHz	112mW	[61]

4.7.1 用于目标追踪的最大值检测法

最大值检测法(MaxDet)是一种通过在包含被追踪物体的图像中检测最大的像素信号值来实现目标追踪的方法。文献[383]中用到了 WTA 电路,通过在每个像素中集成 WTA 电路来检测整幅图下像素信号的最大值。WTA 电路在 3.3.1 节中已详述。为了得到最大值的 x-y 位置,在行与列放置两个一维电阻网络。一个 $62\mu m \times 62\mu m$ 的像素包括 WTA 电路和一个寄生光电晶体管。芯片的像素阵列为 24×24,像素的处理速度达到 7000pixel/s。另一种运用最大值检测法的方法是在行与列的方向各放置一个一维的电阻网络,通过与比较器一起工作,得到每个方向最大值的位置[384]。

4.7.2 用于目标追踪的投影法

已经有研究使用投影法来进行目标追踪[173,385-387]。因为只需要沿像素阵列行和列的求和操作,所以投影法很容易实现。沿每一行和列方向上的投影是一阶图像矩,通过进行数据预处理,如边缘检测[385]和二值化[386],可以有效得到矩心。文献[385]中,成像区域被分成了两部分:中央区域与周围区域,两部分的像素密度是不一样的。该结构模仿的是感光体在人的视网膜的分布。拥有密集像素密度的中央区的功能是完成边缘检测和运动检测,而拥有稀疏像素密度的周围区域的功能是完成边缘检测和投影。这种光电探测器结构类似于 3.2.2 节介绍的对数传感器。对数传感器含有一个光电晶体管和一个亚阈值晶体管,可以实现对数响应。边缘检测通过一个电阻网络来实现。

研究者研制出了一种智能 CMOS 图像传感器,它可以同时完成普通的成像以及在行与列上的投影[387]。在这种传感器中,$20\mu m \times 20\mu m$ 的像素里拥有 3 个 FD:一个完成成像,另外两个完成投影工作。该设计中有源像素完成成像,无源像素通过将某个方向所有像素电荷做加和处理以完成投影。该传感器具有全局曝光及随机访问功能,这对于实现高速成像十分有用。该传感器拥有 512×512 的像素阵列,用 $0.6\mu m$ 标准 CMOS 工艺制成。

4.7.3 基于电阻网络及其他模拟处理的目标追踪

另一种模拟处理是电阻网路技术。图 3.7 实现了一个双层的电阻网络[45]。时间上和空间上的差值操作可以通过帧间做差值来实现。目标追踪在芯片外进行中值滤波。因为采用有源像素和消除电路,这个具有电阻网络的传感器可以获得很高的信噪比以及很低的固定模式噪声。这种传感器的结构在 3.3.3 节中已详述。

另一种基于细胞神经网络(CNN)的全并行模拟处理方法已经提出[388]。像素尺寸为 $85\mu m \times 85\mu m$,基于 $0.18\mu m$ 标准 CMOS 工艺,像素内部集成了模拟信号处理电路及异步数字电路,像素阵列为 64×64。芯片可实现处理的时间为 $500\mu s$,速度与功耗大约为 725MIPS/mW。如果使用全模拟结构的话,在 1.8V 供电的情况下,功耗仅为 $243\mu W$。

4.7.4 基于数字处理的目标追踪

最后我们讨论采用数字处理的芯片。这种芯片采用 $0.5\mu m$ 标准 CMOS 工艺制作了 $80\mu m \times 80\mu m$ 的像素,在每个像素中集成了 1 位的数字处理单元或串行处理单元。图 4.28 给出了芯片的示意图。像素电路已经在图 3.28 中给出。因为实现了全数字信号处理,所以芯片是全部可编程的,可以达到很高的处理速度,大约为 1ms。这样的采用数字处理的芯片需要在每个像素的内部集成一个 ADC,这块芯片采用 3.4.1 节介绍的由反相器构成的类 PWM 型 ADC。

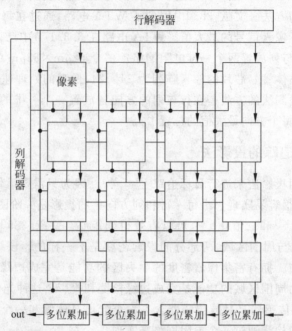

图 4.28　全数字智能 CMOS 图像传感器的示意图[61]（BSCA：位串行累加器）

这种芯片已经被应用在显微镜反馈系统中,其与焦深法相结合成功实现了运动的微生物标本的高速追踪,结果如图 4.29 所示[389]。通过光学显微镜观察运动的物体是比较困难的,通过运用目标捕捉技术可以实现光学显微镜系统对微小的物体进行自动的追踪。文献[389]的实验报告显示使用工作频率为 1kHz 的数字视觉芯片可以得到很好的追踪结果,同时也可以结合焦深技术实现三维的追踪。

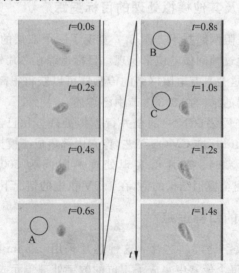

图 4.29　移动的生物标本的追踪结果。照片由安装在光学显微镜上的 CCD 相机拍摄。追踪通过全数字视觉芯片实现[389]（在此向东京大学的 M.Ishikawa 教授致谢）

4.8 像素与光学系统的专用装置

本节讨论拥有专用像素阵列或光学系统结构的智能 CMOS 图像传感器。传统的智能 CMOS 图像传感器用一个光学透镜将图像聚焦到传感器的成像平面上,在成像平面上有矩阵式正交排列的像素。另外在一些视觉系统中也采用非正交排列的像素阵列进行成像。一个很典型的例子就是人类的视觉系统,眼睛的光感受器的分布是不均匀的:在中央区(或者叫做视网膜中心凹)的像素密度排列紧密,而在周围区像素密度较为稀疏[187]。这种结构在某些情况下很有用,因为它可以在一个宽视角中快速地检测到物体。一旦检测到目标物体,眼睛就会直接指向目标物体并用视网膜中心区域进行更精确的成像。另一个例子就是昆虫的复眼[390,391],在昆虫的复眼中,它的复杂的像素排列是与特殊的光学系统相结合的。CMOS 图像传感器的像素排列方式可以做到比 CCD 图像传感器更复杂,因为对于 CCD 图像传感器来说,像素之间的准直排列对电荷的传输有着关键性的影响。一个弯曲排列的像素阵列会造成 CCD 图像传感器电荷转移效率的大幅降低。

此节首先讨论具有特殊像素阵列的智能 CMOS 图像传感器,然后介绍具有特殊光学系统的智能 CMOS 图像传感器。

4.8.1 非正交排列

4.8.1.1 类似视网膜中心凹的传感器

如图 4.30 中的类似视网膜中心凹的传感器模仿的是人眼视网膜结构。在人的视网膜中,中央区的光感受器的密度要比外围区大[392]。

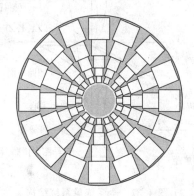

图 4.30 类似视网膜中心凹的传感器的像素排列,像素尺寸以平方根率的形式增长,即像素间距按照对数率降低,中间区域为控制电路

类似视网膜中心凹的图像传感器拥有类似于人眼视网膜结构的像素排列:像素的密度沿着径向方向按对数式的降低,称为级数坐标系[393-396]。在对级数坐标系中,直角坐标系通过下式转换为极坐标系:

$$r = \sqrt{x^2 + y^2}, \quad \theta = \arctan\left(\frac{y}{x}\right) \qquad (4.26)$$

这种类似视网膜中心凹(或者称为对级数的图像传感器)对于一些图像的处理很有用

处,但是在版图设计上很困难。首先,由于中央区域的像素密度很高,在中心位置放置像素很难,所以在中心区域内很可能无法完成成像。其次,径向与环形的两个扫描器很难布局。为了解决这些难题,R. Etienne-Cummings 等人研究了一种只有两类区域(中央区、周围区)的类视网膜中心凹的传感器,每个区域内部的像素阵列仍然使用正交排列或传统排列[385]。这种传感器可以用于目标追踪,详见 4.7 节。扫描器布局问题在全景成像中依然存在。

4.8.1.2 全景成像

全景成像(HOVI)是一种可以拍摄周围所有方向上的图像的成像系统,它可以通过一个传统的 CCD 相机与一个双曲面镜来实现[379,397],它很适合用于监视。输出图像是被反射镜投射得到的,所以会存在失真。通常情况下,失真图像被变换到直角坐标系,进行重新排列,然后显示出来。这样的相机外变换操作限制了它的应用。CMOS 图像传感器在像素排列方面具有多样性。因而,像素的排列可以经过设计以适应被双曲面镜扭曲的图像。这样可以实现直接的影像输出,无须任何软件转换程序,从而使传感器有更广泛的应用。本节中,我们会讨论一个用于全景成像的智能 CMOS 图像传感器的结构和特征[398]。

一个传统的全景成像系统通常包括一个双曲面镜、透镜和 CCD 相机。全景成像系统拍摄的图像由于双曲面镜的原因会发生失真。为获得一个可辨认的图像,必须经过变换过程,这通常由计算机软件来实现。图 4.31 给出了 HOVI 成像的原理。

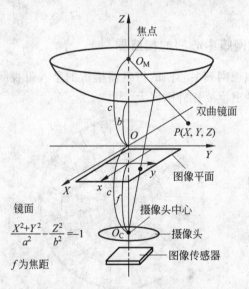

图 4.31 全景成像系统结构

一个位于 $P(X,Y,Z)$ 的物体通过双曲面镜投射到一个二维图像上,坐标为 $p(x,y)$。$p(x,y)$ 的坐标由下式得到

$$x = \frac{Xf(b^2 - c^2)}{(b^2 - c^2)Z - 2bd\sqrt{X^2 + Y^2 + Z^2}} \tag{4.27}$$

$$y = \frac{Yf(b^2 - c^2)}{(b^2 - c^2)Z - 2bd\sqrt{X^2 + Y^2 + Z^2}} \tag{4.28}$$

式中,b 和 c 是双曲面镜的参数,f 是相机的焦距。

图4.33(a)展示了使用CCD相机的全景成像系统获得的图像信息。输出图像是失真的。根据式(4.27)和式(4.28),我们可以将智能CMOS图像传感器的像素设计成放射状阵列[398]。传感器用0.6μm、三层金属、两层多晶硅的标准CMOS工艺制造。制造芯片的参数如表4.4所示。

表4.4　HOVI的智能CMOS图像传感器的规格

工艺	0.6μm、两层多晶硅、三层金属标准CMOS
芯片尺寸	8.9μm * 8.9μm
像素数	32×32
PD结构	n扩散/p衬底
电源电压	5V
PD尺寸	沿着径向从第1到第8个像素 18μm * 18μm
	从第9到第16　20μm * 20μm
	从第17到第24　30μm * 30μm
	从第25到第32　40μm * 40μm

像素电路使用3T有源像素结构。芯片的一个特征是像素的间距由外围到中心越来越小。所以如表4.4所示,该传感器使用了四种不同尺寸的像素。在放射状结构中,垂直和水平扫描线分别被放置到沿径向和圆周方向。图4.32给出了所制造的芯片的显微照片。在图中,也给出了最中心像素的特写图。

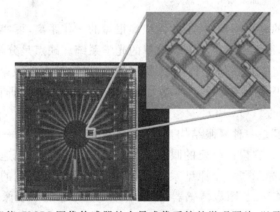

图4.32　基于智能CMOS图像传感器的全景成像系统的微观照片,以及中央像素的特写

图4.33显示了传感器的实验结果。输入图案是一张由基于CCD图像传感器的全方位视觉系统拍摄的图片。结果显示所制造的CMOS图像传感器可以很好地将全方位视觉系统拍摄的图像还原。

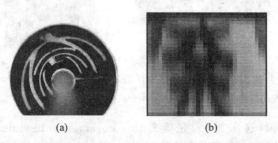

(a)　　　　　　　(b)

图4.33　一个传统的全景成像系统拍摄的日文图像和使用上述CMOS图像传感器的输出图像

4.8.2 专用光学系统

4.8.2.1 复眼结构

复眼是包括昆虫以及甲壳类动物在内的节肢动物的生物视觉系统。复眼中有大量的独立的具有小视场的微小光学系统,具体结构如图4.34所示。每个独立的微小眼睛(称为小眼或单眼)所获得的图像在大脑内组成整个图像。

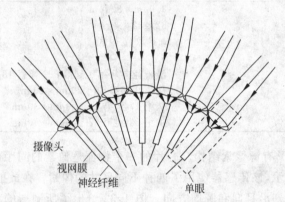

图4.34 生物复眼系统的概念。系统包括大量的小眼,每个小眼拥有一个透镜、视网膜和神经纤维。另一种复眼是神经叠加复眼[390]

复眼的优点是它的宽视场角、紧凑的体积和很短的工作距离,可实现超薄相机系统;另外由于小眼的小视场,每个小眼只需要简单的光学系统。缺点是分辨率较低。

目前已经有很多组织正在研究仿生复眼[399-404]。在接下来的段落中,会介绍两个基于复眼结构的智能CMOS图像传感器:约克大学的R. Hornsey等人研制的"蜻蜓眼",大阪大学J. Tanida等人研制的"TOMBO"。

蜻蜓眼。"蜻蜓眼"是一种复眼结构,它拥有多达20个小眼,在每个小眼的成像平面上,大约有150个像素[404]。它模仿蜻蜓的眼睛系统,并且希望应用于高速的物体追踪以及深度感知等领域。图4.35显示了一个原型系统,此视觉系统可以实现宽的视角。通过使用具有随机访问功能的智能CMOS图像传感器,可以实现对每个小的子成像区域的快速访问。

(a) (b)

图4.35 蜻蜓眼的一个原型系统图:(a)复眼束的特写图;(b)总系统。每个小眼拥有一个透镜和一束光纤(在此向约克大学的R. Hornsey教授致谢)

TOMBO。TOMBO这几个字母是英文thin observation module by bound optics的首字母缩写,是另外一种复眼系统[405,406]。图4.36给出了TOMBO系统的概念。TOMBO系

统的核心引入了大量的光学成像系统,每个光学成像系统中有若干个微透镜,也称为光学成像单元。每个光学成像单元用不同的摄影角度拍摄小的完整图像。因此,我们可以获得大量的拥有不同拍摄角度的微小图像。一个整体图像可以通过将每个光学成像单元图像整合重建来获得。数字的后端处理算法可以提高合成图像的质量。

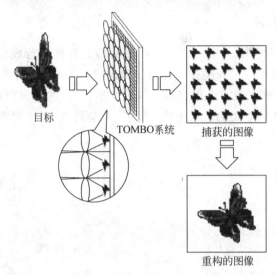

图 4.36　TOMBO 的概念(在此向大阪大学的 J.Tanida 教授致谢)

实现复眼系统的关键问题是用微光学技术设计小眼结构。在 TOMBO 系统中,图 4.36 所示的信号分离器解决了这个问题。专用于 TOMBO 系统的 CMOS 图像传感器已经开发出来了[118,407]。

TOMBO 系统也可以用来作为广角成像系统和超薄型或紧凑型摄像系统。在此系统中,每 3×3 个单元共用一个透镜。图 4.37(a)给出了该系统结构和该系统拍摄的 3×3 单

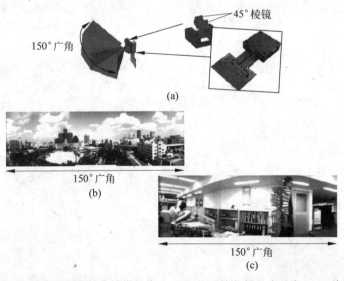

图 4.37　基于 TOMBO 系统的宽视角摄像系统。(a)相机系统包括一个具有 3×3 单元的 TOMBO 相机和两组带有附件夹的棱镜;(b)获得的宽视角图;(c)运动物体的检测(在此向船井电机公司的 Toyoda 先生和 Masaki 先生敬谢)

元图像。通过将两个棱镜拼接,可以得到150°宽视角的图像,如图4.37(b)所示。每个光学成像单元都有一个小的扫描区域,所以采用2.1.1节所述的滚筒曝光并不会引起图像严重失真。如果每个成像单元在不同的时间进行拍摄,就能实现运动物体的检测,如图4.37(c)所示。

4.8.2.2 偏振成像

偏振是光的一种特性[391]。偏振成像利用光的偏振特性进行成像,并且应用在侦测物体需要更高清晰度的领域中。人类并不能感知偏振,然而很多动物可以。通过将双折射材料涂在传感器表面,几种偏振图像传感器已被研发出[408]。这种偏振智能CMOS图像传感器可能在化学领域有比较重要的应用价值,因为在化学研究中,经常要用偏振的形式来确定化学物质的类型。

第5章 应 用

5.1 引言

本章介绍了智能 CMOS 图像传感器的几类应用。

第一类是在包括人机界面在内的通信和信息领域的应用,在这些领域引入成像功能不仅可以提高性能(如通信性能),还可以提升便捷性(如视觉辅助控制)。

第二类是在生物技术领域的应用。在该领域中,广泛使用了结合光学显微镜的 CCD。假如我们使用 CMOS 图像传感器,成像系统可以变得更加紧凑,也可集成更多的功能,这样就能够提高其性能。

最后,是在医学领域的应用。因为成像系统可能会通过吞咽或植入的方式进入人体,所以应用在医学上的系统必须非常紧凑。智能 CMOS 图像传感器以其体积小、功耗低和集成功能多等优点更适用于这类应用。

5.2 信息和通信应用

光源在蓝色 LED 和白色 LED 出现之后发生了极大的变化,例如,一些室内照明灯、汽车灯和大型室外显示屏都使用 LED。可快速调制 LED 的引入产生了一种新的应用——自由空间光通信。由于对人来说图像更加直观,所以结合图像传感器后自由空间光通信可以拓展到人机交互。

在本节中,我们介绍了智能 CMOS 图像传感器在通信和信息方面的应用。首先介绍光学识别标签,接下来介绍无线光通信。

5.2.1 光学识别标签

大型户外 LED 显示屏以及白色 LED 灯等的普及促进了 LED 应用到空间光通信系统接收机中的研发。可见光通信就是这样一个系统[409],这项应用在5.2.2节进行了描述。另一种应用是使用 LED 作为一个高速调制器来发送 ID 信号的光学识别标签。商业化的光学识别标签已经用来进行动作获取[410,411]。业界已经提出并开始研发在增强真实性技术上应用光学识别标签的系统,如下面的几个系统:Phicon[412]、Navi cam[413]、ID cam[414,415]、A*i*mulet[416,417] 和 OptNavi[418-420]。

Phicon,A*i*mulet。这些系统已经被用于为用户发送数据并接收来自用户的数据。用户有一个终端来记录他的位置,同时从基站发送/接收数据。Phicon 系统由 GITech 的 D. J. Moore 和 Xerox PARC 的 R. Want 等人提出[412]。该系统使用一个红外 LED 灯作为光学信号柱来传递用户的位置和发送数据。一个单色 CCD 用来定位光学信号柱和解码红外收发器传送的数据。由于该系统采用了传统 CCD 摄像头进行调制光的解码,所以比特率大约是

8bit/s。这项数据表明,为满足更高速的数据率探测信号柱的要求,必须用专用的相机系统。

A*i*mulet 是由日本的 H. Itoh 在 AIST(Advanced Institute of Science and Technology)开发出来的。该系统作为一个便携式的通信设备,可以让用户在博物馆和展览品中寻找信息。一些演示模型已经被开发并在日本大型的爱知世博会上进行了测试。

ID 相机,OptNavi 系统。ID 相机是由索尼的 Matsushita 等人提出的,该相机使用一个有源的 LED 源作为光学信号柱,该光学信号柱带有一个解码传送识别数据的专用智能 CMOS 图像传感器,该智能传感器将在后面描述。

OptNavi 是由日本 NAIST(Nara Institute of Science and Technology)的创作小组提出的,该系统是为手机和 ID 相机设计的。如图 5.1 所示,这些系统的一个典型应用是大型 LED 显示屏上的 LED 也可以作为光学信号柱。图 5.1 中,LED 显示屏发送 ID 数据,当用户用能解码这些识别数据的智能 CMOS 图像传感器拍摄照片时,解码数据会附加在用户界面上,用户很容易得到显示内容的相关信息。

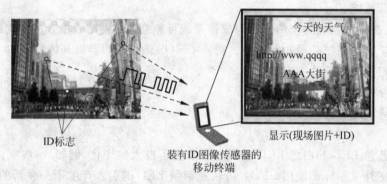

图 5.1 光学识别标签的概念。大型 LED 显示屏上的 LED 被用作光学识别标签。识别数据附加在用户拍摄的图像上

另一个应用是作为一个连接在网络中的电子设备的可视化远程遥控器。人们对由电子设备构建的家庭网络越来越感兴趣,并且已经成立了一些论坛,如 DLNA(Digital Living Network Alliance)[421]、ECHONET(Energy Conservation and Homecare Network)[422]、MPnP™(Mniversal Plug and Play)[423] 和 HAVi(Home Audio Video Interoperabitily)[424] 等。人们可以通过家庭网络连接许多网络家电。OptNavi 系统是作为可视化地控制联网家电的人机界面而专门提出来的。在这个系统中,装有定制图像传感器并配备了大屏幕、数码摄像头、红外数据通信(IrDA)[425]、蓝牙(Bluetooth)[426] 等的手机作为接口。在 OptNavi 系统中,家庭网络中的电器(如电视、DVD 刻录机和电脑等),都配备了在 500Hz 的频率下传输 ID 信号的 LED。具有多个感兴趣区域(ROI)高速读出的图像传感器用作 ID 信号接收装置。如图 5.2 所示,接收到的识别信号叠加在传感器捕获的背景图像上。通过使用 OptNavi 系统,我们可以在手机显示屏上可视化地控制这些家电。

用于光学识别标签的智能 CMOS 图像传感器

由于传统的图像传感器以 30 帧/秒的速率捕捉图像,所以它不能收到 kHz 速率的光学识别信号。因此,可接受识别信号对于自定义图像传感器来说是非常必要的。如表 5.1 所

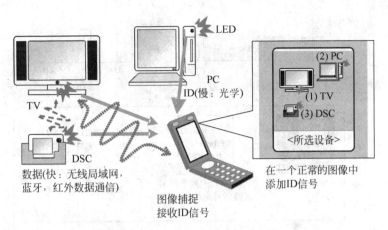

图 5.2　光学导航的概念。专用的智能 CMOS 图像传感器检测和解码家用电器的光学识别标签。解码的结果叠加在传感器采集的图像上

示,目前已经提出并验证了专用于光学识别标签的智能 CMOS 图像传感器,这些传感器以较高的帧频速率接收识别标签信号。索尼的 S. Yoshimura 等人和来自同一团队的 T. Sugiyama 等人证实了图像传感器的所有高速读出操作的像素和传统的 CMOS 图像传感器具有相同的像素架构,这些传感器都非常适合高的分辨率。

东京大学的 Oike 等人已经证实了通过以传统的帧频速率捕捉图像的传感器可以接收高精度的识别标签信号,该传感器通过像素中的模拟电路来接收识别标签信号。

表 5.1　用于光学识别标签的智能 CMOS 图像传感器的规格

机构	索尼	东京大学	NAIST
来源	[375]	[427]	[431]
ID 检查	高速读出像素	像素内 ID 接收	多个 ROI 读出
工艺	$0.35\mu m$	$0.35\mu m$	$0.35\mu m$
像素尺寸	$11.2\mu m \times 11.2\mu m$	$26\mu m \times 26\mu m$	$7.5\mu m \times 7.5\mu m$
像素数目	320×240	128×128	320×240
特征频率	14.2kfps	80kfps	1.2kfps
功耗	82mW(@3.3kfps,3.3V)	682mW(@4.2V)	3.6mW(@3.3V, w/o ADC, TG)
特点	功耗低	大尺寸	速度快

低功耗多个 ROI 的高速读出电路。所有像素的高速读出电路都会引起大的功率消耗。NAIST 团队已经提出了一种专用于光学识别标签的图像传感器,它可以利用单像素电路实现低功耗的高速读出。在读出方案中,传感器以传统的视频帧频速率捕捉正常的图像,同时捕捉多个 ROI,它可以以高速的帧频速率接收识别标签信号。为了定位传感器中的 ROI,引入了一个 10Hz 的低频光学控制信号,如图 5.3 所示,传感器可以利用帧差分法很容易地识别这个信号。

如图 5.4 所示的读出方案,其特征是基于多个 ROI 的交叉读出电路,每个 ROI 在正常图像的一帧中会被多次读出,所以 ROI 的读出速率比正常图像的帧频速率快得多。为了使读出方案解释简单,图 5.4 中显示了 6×6 的像素,因为其中存在两个识别标签,所以显示两个 ROI,在图中,每个像素中的数字表示读出顺序,括号中的数字表示参与识别标签信号的

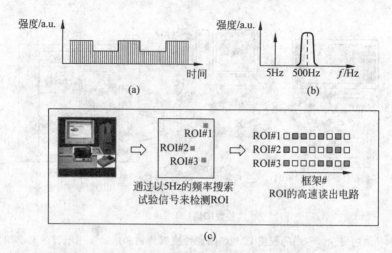

图 5.3 使用低频的控制信号检测 ID 位置的方法:(a)控制信号和识别标签信号的波形;(b)控制信号和识别标签信号的功率谱;(c)使用控制信号获得识别标签信号的过程

1	2	3	4	7	8	9	10
13	14	15	16	19	20	21	22
25	26	27	28	31	32	33	34
37	38	(5), (29), 39, (53)	(6), (30), 40, (54)	43	44	45	46
49	50	(11), (35), 51, (59)	(12), (36), 52, (60)	55	(17), (41), 56, (65)	(18), (42), 57, (66)	58
61	62	63	64	67	(23), (47), 68, (71)	(24), (48), 69, (72)	70

图 5.4 只画出了 6×6 像素的多个 ROI 的快速读出说明

像素。在本例中,整幅图像的一帧期间读出了 3 个 ROI。

传感器中的 ROI 的帧频速率是 1.1kfps。像素的数量是 320×240,ROI 的尺寸是 5×5,识别标签的数量是 7,整幅图像的帧频速率是 30fps。在读出方案中,系统的时钟速率和传统的图像传感器相同,是 60fps,因此即使以很高的帧频速率读 ROI,也仅消耗很低的功耗,通过关闭列或者 ROI 的电源供应可以进一步地降低功耗。

如图 5.5 所示的传感器的框图,为了高速读出 ROI 和切断 ROI 外像素的电源供应,传感器是以识别标签的映射表进行工作的,即以 1 比特位的存储数组形式存储识别标签位置信息。如图 5.5(b)所示,这个像素电路很简单,与传统的 3T-APS 相比,它的列复位只有一个额外的晶体管。如 2.6.1 节所描述的,这个额外的晶体管必须嵌入在复位线与复位晶体管的栅极之间。这个晶体管用 2.6.1 节中所说的复位和随机存储来读出 ROI 中的像素值。图 5.6 显示了这个传感器的时序图。在这个时序图中,正常图像的碎片和 ROI 图像交叉读出。

图 5.7 是传感器显微照片。表 5.2 中总结了性能规范。图 5.8(a)显示了以 30fps 速率捕捉到的正常图像,图 5.8(b)和图 5.8(c)显示了识别检测的实验结果。识别信号从 3 个 LED 模块中以 500Hz 的 8 位微分编码调制进行传输。每个识别信号有 36 帧的 ROI 图像,每帧图像由 5×5 像素构成,如图 5.8(c)所示,当捕捉到整幅图像的一帧后就会检测识别信

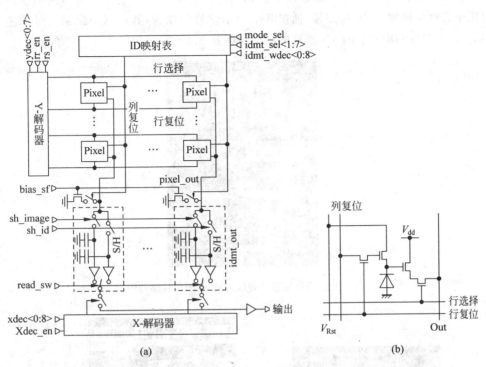

图 5.5 (a)智能 CMOS 图像传感器的多个 ROI 的快速读出框图；(b)像素电路

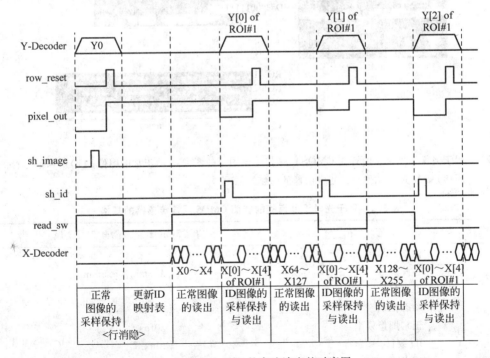

图 5.6 多个 ROI 的高速读出的时序图

号。ROI 图像的这种模式成功地验证了识别信号的探测。这个结果证明图像传感器可以以 30fps 的速率捕捉 QVGA 图像,捕捉 3 个识别信号的图像,其中每个识别信号的速率是 1.1kfps。3.3V 电压供应下的传感器的功率消耗是 3.6mW。

图 5.7　用于光学识别标签的智能 CMOS 图像传感器的显微照片

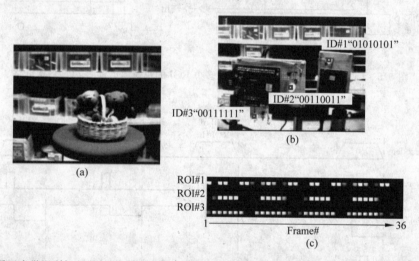

图 5.8　用于光学识别标签的智能 CMOS 图像传感器拍摄的图像:(a)正常的图像;(b)识别信号检测; (c)36 帧的 ROI 图像/ID(正常图像的一帧)

表 5.2　用于光学识别标签的智能 CMOS 图像传感器的规格

工艺	0.35μm(2 层多晶硅 3 层金属)
像素数目	320×240
芯片尺寸	4.2mm×5.9mm
像素尺寸	7.5μm×7.5μm
动态范围	54dB
正常图像频率	30fps
ID 数目	Max. 7
ID 图像频率	1.2kfps/ID
功耗	3.6mW@3.3V

5.2.2 无线光通信

在这一节中,介绍使用智能 CMOS 图像传感器进行无线光通信或者自由空间光通信的基本知识。无线光通信比传统的光纤通信和射频优势大,主要有以下几个原因。首先,建立一个无线光通信只需要很小的投资,这表明,无线光通信可以应用于多个建筑物之间的通信。其次,它在 Gbps 的高速通信领域也有很大的潜能。再次,它受其他电子设备的干扰很小,这对于植入电子医疗设备来说是非常重要的。最后,因为它的多样性很窄,所以它是安全的。

从应用方面来分析,可以将无线光通信分为三类。第一个系统是用在户外或者超过 10m 的长距离的无线光通信。主要的应用目标是建筑内的局域网,光速可以很容易地以很高的数据速率连接建筑中的两点。工厂中的局域网是这个系统的另外一个应用,低电磁干扰(EMI)和安装方便的特点非常适合在嘈杂环境下的工厂中使用。这些应用的很多产品已经商用化了。

第二个系统是近距离光通信。红外数据通信及与其相关的通信都属于这一类。由于 LED 速率的限制[432],数据速率并不是很快。这种系统可以与射频无线通信(如蓝牙)相媲美。

第三个系统是与第二个系统相似的室内光无线通信,但是主要应用于局域网,至少有 10MHz 以上的数据速率。这样的室内光无线系统已经商用化了,但是十分有限[433]。图 5.9 是一个在室内使用的光无线通信系统。这个系统由安装在天花板上的枢纽站和位于计算机旁边的几个节点站组成。它是一对多的通信架构,而室外的多是一对一的通信架构,这种一对多的架构容易引起几个问题。为了使节点能够找到枢纽中心,枢纽中心需要配备一个可移动的、笨重的光学器件机械装置(如光接收器),这就意味着节点站的规模相对较大。但是,由于系统是在室内使用,所以中心和节点的尺寸是至关重要的。

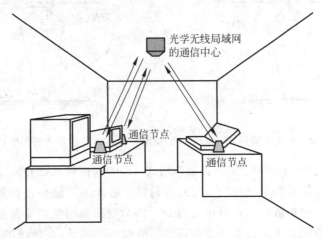

图 5.9 室内光无线通信系统的例子。天花板上安装了一个枢纽站,其他的几个计算机附近的节点与枢纽站进行通信

蓝色和白色 LED 的快速发展,使一种被称为可见光通信的新型室内光无线通信成为可能[409]。许多使用这一概念的应用已经被提出并开发,例如,室内的白色 LED 灯可被用作收

发器,汽车上的 LED 可以用来与其他汽车进行通信。这些应用都与 5.2.1 节中的光学识别标签紧密相关。

在光无线通信中,利用二维探测阵列可以增强接收效率[434,435]。用智能 CMOS 图像传感器取代这种二维探测阵列进行快速光无线通信有很多的优势。NAIST[436-446]和加州伯克利分校[447]已经提出并验证了这种传感器。加州伯克利分校正在研究小型无人机之间的通信[447,448],即相对室内通信来说具有较长的室外通信距离。

接下来,将介绍一种室内光无线局域网的新方案并进行详细描述,其中将会用到一种基于光接收器的图像传感器。

使用智能 CMOS 图像传感器的光无线局域网。在室内光无线通信的新方案中,智能 CMOS 图像传感器被用作光接收器,这与探测中心和节点的通信模块位置的二维位置感应装置一样。相比之下,在传统的系统中,一个或者多个光电二极管被用来探测光信号。

图 5.10 对比了所提出的室内光无线局域网和传统的系统。在传统的光无线局域网中,光学信号必须准确地向对向传输,从而实现以特定的光学输入功率入射到光学探测器的功能,因此探测和对齐光的通信模块位置非常重要。如图 5.10(a)所示,在自动节点探测的光无线局域网系统中,用于光电探测的机械式扫描系统在节点中用来寻找中心。然而,由于聚焦透镜必须聚焦足够的光功率,所有透镜的直径都很大,这使得扫描系统的体积很笨重。另一方面,如图 5.10(b)所示,使用一个图像传感器作为光接收器。这种方法有几个非常好的光无线通信的特征。因为图像传感器可以通过简单的图像识别算法(而不用机械器件)很容易地捕捉通信模块的周围环境,从而指定其他模块的位置。此外,图像传感器固有的可以通过大量的微二极管并行地捕捉多个光信号。所以,当图像传感器有足够的空间分辨率时,独立的像素可以探测不同的模块。

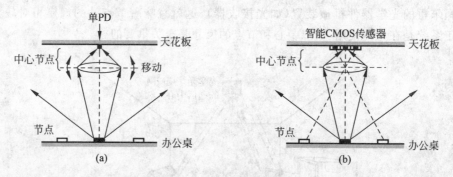

图 5.10 (a)传统的室内光无线局域网系统;(b)使用智能 CMOS 图像传感器的光无线局域网系统

这种传感器有两种工作模式:图像感应模式(IS)和通信模式(COM)。如图 5.11(a)所示,枢纽中心和节点工作在 IS 模式下,它们传输和扩散光来互通存在。为了尽可能覆盖对照物存在的区域,采用扩散光的辐射角度为 2θ。因为对照物的光功率探测信号通常比较微弱,所以在图像感应模式下进行探测非常有效。值得一提的是,在传统的系统中,光收发器需要通过摆动笨重的光学仪器扫描房间里其他的收发器,这既耗时又浪费能量。

如图 5.11(b)所示,在 IS 模式下指定位置后,中心和节点的传感器工作模式转变为通信(COM)模式。它们向对照物的方向发出带有通信数据的窄的平行光束。在通信模式中,光电流直接读出,而不需要经过接收光信号的特定像素的时间整合。在通信模块和接收电路区域中使用平行光束能够降低功耗,这是因为发射极的输出功耗和光接收器的增益减少了。

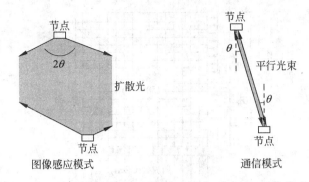

图 5.11 传感器的两种模式:(a)图像感应模式(IS);(b)通信模式(COM)

使用智能 CMOS 图像传感器的系统还有一个好处,它们可以采用空分复用(SDM)增加通信带宽。图 5.12 显示了系统中的 SDM。当来自不同通信模块的光信号被独立的像素接收之后,在读出行中分开读出,这样从多个模块中进行并行数据采集才能实现。因此,中心下行线的总带宽随读出行数的增加而成比例增加。

为了读出通信模式下的光电流,如图 5.12(b)所示,可以使用一个所谓的集中读出。因为聚集光斑的大小是有限的,所以来自通信模块的光信号被一个或者几个像素接收到。来自接收光信号像素的放大的光电流和为通信模式准备好的读出行是相同的,所以,信号电平并不会减小。

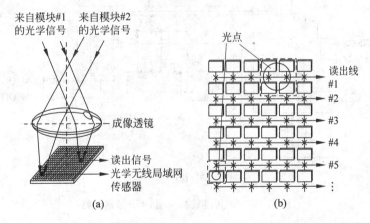

图 5.12 使用集中读出的传感器进行空分复用的示意图:(a)多光斑的读出;(b)传感器上的集中读出的原理图

智能 CMOS 图像传感器的实现。如图 5.13(a)所示为传感器的框图。传感器有一个图像输出和四个数据输出,图像是通过采样保持电路输出的。芯片在通信模式下有四个数据输出通道。图 5.13(b)显示了像素电路,该电路包括一个 3T-APS、一个跨阻放大器(TIA)、一个模式切换数字电路和一个锁存器。为了能够切换 COM/IS 模式,可向锁存器写入一个高/低信号。跨阻放大器的输出转换为一个电流信号并与相邻像素的信号取和。如图 5.13(b)所示,每个像素分布在左右两边数据输出线,并与上述信号集中读出。经过跨阻放大器转换和主放大器的放大,电流信号之和输入到行列中。传感器采用标准的 0.35μm CMOS 工艺制作。规格总结在表 5.3 中,制作芯片的缩影照片如图 5.14 所示。

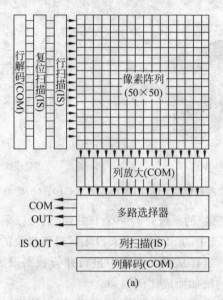

(a)

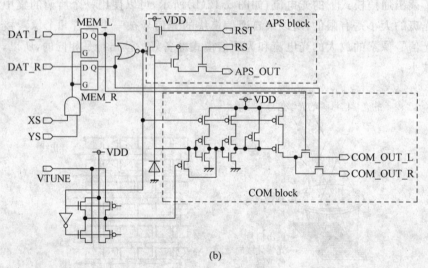

(b)

图 5.13 (a)传感器的框图;(b)像素电路

表 5.3 室内光无线局域网所采用的智能 CMOS 图像传感器的规格

工艺	0.35μm(2 层多晶硅 3 层金属)
像素数目	50×50
芯片尺寸	4.9mm×4.9mm
像素尺寸	60μm×60μm
二极管	n 阱 p 衬底
填充因子	16%

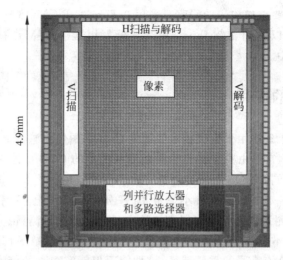

图 5.14 室内光无线局域网所采用的智能 CMOS 图像传感器的显微照片

图 5.15 显示了使用制作的传感器进行成像和通信的实验结果。830nm 波长的光的感光灵敏度是 70V/(s×mW),接收到的波形显示在图 5.15 中。眼睛在 650nm 条件下可以达到 50Mbps,在 830nm 条件下达到 30Mbps。在该传感器中,使一束强烈的激光入射到一些像素上从而实现快速通信,同时其他的像素工作在图像感应模式下。激光束入射到传感器上产生大量的光生载流子,它们依据波长移动很长的距离,一些光生载流子进入光电二极管而影响图像。图 5.16 显示了两个波长下,传感器的有效扩散长度的实验结果。正如 2.2.2 节所述,波长越长,扩散距离越长。

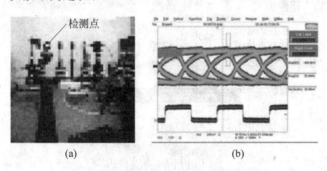

图 5.15 智能 CMOS 图像传感器的实验结果:(a)捕捉的图像;(b)830nm 波长下 30Mbps 眼睛模式

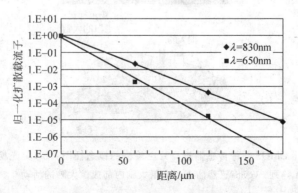

图 5.16 像素中扩散载流子的实测数量

提高数据速率需要更加深入的研究。采用 0.8μm 的 BiCMOS 技术的初步结果显示每个通道的数据速率可以达到 400Mbps。引入波分复用器（WD-M）的可提高数据速率的系统[444-446]已经在研究中。

5.3 生物技术的应用

在这一节中，将会介绍用于生物技术的智能 CMOS 图像传感器。荧光检测是一种广泛使用的生物测量技术，它是由安装在光学显微镜系统上的 CCD 摄像头进行测量的，这已经被确定为一个采用智能 CMOS 图像传感器有效成型的应用。将智能 CMOS 图像传感器引入生物技术可以带来集成功能和小型化的好处，结合片上探测这些优势会得到增强[278,449-453]。片上探测意味着标本可以直接放在芯片表面进行测量。这样的配置使它能够很容易的直接访问标本，所以荧光、电势、pH[452]和电化学的参数等都能够测量。集成性能也是智能 CMOS 图像传感器的重要特征，该性能不仅能实现高的信噪比，也能实现功能上的突破。例如，将电模拟集成到传感器中，使测量细胞刺激引起的荧光成为可能，这是一个片上电-生理测量方法。传感系统的总尺寸能够通过小型化而减小，这种小型化使植入传感系统和球场上的测量都成为可能，这是因为系统的总尺寸减小增强了系统的移动性。

图 5.17 显示了使用智能 CMOS 传感器片上探测的典型例子。在图 5.17(a)中，实现了对神经元的染色，当神经元收到外部刺激而兴奋时会显示出荧光。由于传感器有片上电子模拟器，可以刺激神经元和检测相关的荧光。图 5.17(b)显示使用智能 CMOS 图像传感器对 DNA 进行鉴定。DNA 鉴定是通过 DNA 阵列和荧光实现的。单股的 DNA 固定在传感器的表面，目标 DNA 与荧光染料杂交互补链也在传感器表面。如图 5.17(c)所示，第三个

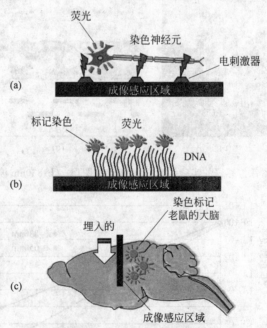

图 5.17 使用智能 CMOS 图像传感器进行片上检测的概念插图：(a)荧光检测到神经的活动；(b)DNA 阵列；(c)传感器插入到老鼠大脑内部成像大脑的活动

例子是在老鼠的大脑中植入智能 CMOS 传感器。此种情况下,传感器很小,可以插入一只老鼠的大脑中,在大脑内部,传感器可以检测荧光以及刺激周围的神经元。有些情况下,荧光强度很弱,需要低光成像技术,例如脉冲调制(详见 3.4 节)[199,200] 和 CMOS 技术构成的 APD 阵列(详见 2.3.4 节)[454-456]。

本节介绍了两个体现上述优势的示例系统:一个是能够将静电成像或电化学成像表现得和光学成像一样的多模式传感器,另一个是体内图像传感器。

5.3.1 多模式功能的智能图像传感器

在这一节中,将介绍多模式功能的智能 CMOS 图像传感器。多模式功能对生物技术尤其有效。例如,在 DNA 鉴定中把光学图像和其他的物理值(如电势图)结合起来,DNA 鉴定就会变得更加准确。在这里介绍了两个例子:光-电势多重成像和光-电化学成像。

光学和电势成像的智能图像传感器

传感器的设计。图 5.18 显示了制作的传感器的显微照片。传感器有一个 QCIF(176×144)的像素阵列,其中包括 88×144 的光学感应像素和 88×144 的电势感应像素,像素的尺寸是 $7.5\mu m \times 7.5\mu m$。传感器由 $0.35\mu m$(2 层多晶硅 4 层金属)的标准 CMOS 工艺制造。

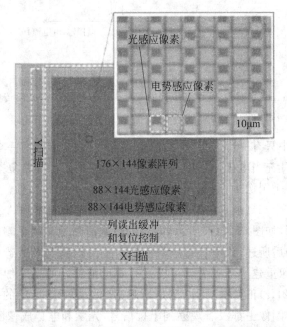

图 5.18 光学与电势双成像的智能 CMOS 图像传感器的显微图像

图 5.19 显示光感应像素、电势感应像素和列单元的电路。电势感应像素由感应电极、源极跟随器和选通晶体管构成。感应电极由一个顶层金属层和覆盖着一层氮化硅的钝化层(LSI)构成。感应电极与芯片表面电势是电容耦合。使用电容耦合的测量方法时,图像传感器没有电流,测量引起的扰动比电导耦合感应传感器系统引起的小得多,如多重电极阵列。

图 5.20 显示了捕获(光学和电势)的图像和重建图像的实验结果。该传感器是硅橡胶

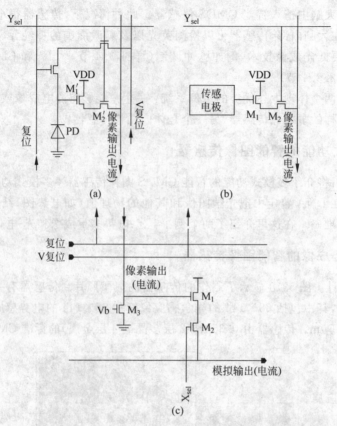

图 5.19 智能 CMOS 图像传感器,光学和电势的双重成像的电路:(a)光学传感像素;(b)电势传感像素;(c)一列单元

模型,在传感器阵列中只有一部分浸入生理盐水溶液。该盐溶液由一个电压源控制,由浸没在溶液中的 Ag/AgCl 充当电极。如图 5.20 所示,拍摄的图像非常复杂,这是由于光学的和电势的图像叠加在一个图像中,数据可以被划分成两个不同的图像。在图 5.20(a)的显微照片中可以观察到灰尘和划痕;在图 5.20(b)中可以清楚地观察到 Ag/AgCl 电极的阴影的光学图像。另一方面,如图 5.20(c)所示,电势的图像清晰地显示了曝光区域的电势与图中覆盖区域的对比。如后所述,该电势感应像素显示了由检测电极所造成捕获电荷相关像素的失调。不过,在图像重建过程中失调可以被有效地消除。如果施加于生理盐水溶液中的电压为 0~3.3V(0.2Hz)正弦波,可以清楚地观察到暴露于溶液中的那部分区域的电势变化。电势图像和光学图像中都没有观察到串扰信号,光学和电势成像已成功实现。

图 5.21 显示了使用导电凝胶探头的电势成像效果。两个独立进行电压控制的导电凝胶探针被放置在传感器上。该图像清楚地显示施加在凝胶点上的电压,结果显示外加电压差较大的容易被捕获,该传感器不仅能够拍摄静止图像,而且能够拍摄 1~10fps 范围内的运动图像。下一个版本的传感器将对帧速率进行改善。通常情况下分辨率都是小于 6.6mV 的,但这足以检测 DNA 杂化[464]。预计电势分辨率将提高到 10μV 的水平。由于目前传感器不具有片上模数转换器(ADC),数据受到从传感器到 ADC 芯片间的信号线的噪声干扰。用于高分辨率神经录制的图像传感器需要使用片上 ADC 电路。

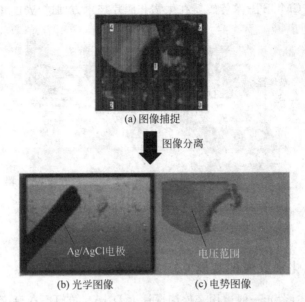

图 5.20 传感器拍摄的图像：(a)拍摄图像；(b)由光学图像(a)重建的图像；(c)从图像(a)重构的电势图像

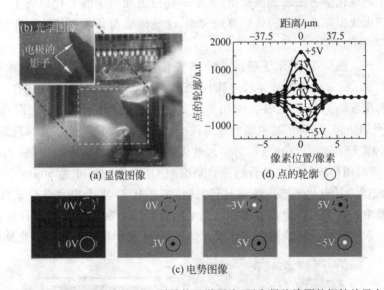

图 5.21 电势成像的实验装置和结果：(a)测量的显微照片，两个凝胶涂覆的探针放置在传感器表面上，并加上电压；(b)片上光成像；(c)片上电势成像；(d)(c)中实心圆表示的点的分布

5.3.2 结合 MEMS 技术电位成像

P. Abshire 和其同事在马里兰州大学采用 CMOS 技术结合微机电系统（MEMS）技术[465]开发了一种电位成像设备，该装置如图 5.22 所示。用于单独培养细胞的微量样品瓶和小孔制造在集成电位检测电路阵列的 CMOS LSI 芯片表面。MEMS 技术用来制造可控的盖子，可以通过驱动器打开或关闭微量样品瓶。通过使用该设备，将细胞通过介电电泳引入并可以在一个较长的时间段进行测试。该设备是目前正在开发的，并已证明了结合

MEMS 技术的智能 CMOS 图像传感器在生物和医疗技术方面的应用。

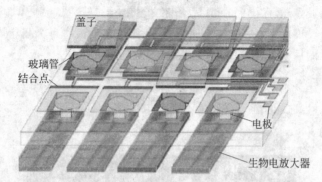

图 5.22　基于生物芯片的 CMOS-MEMS 图示[465]，每一个像素包含一个微量试管和一个感应细胞外电势的放大器(在此向马里兰州大学的 P. Abshire 教授致谢)

5.3.3　光学和电化学成像的智能 CMOS 传感器

如本节开始所述，荧光一般被用来检测探针上 DNA 点的 DNA 杂交靶片段。电化学测量是另一种很有前景的探测机制，它可能成为微阵列技术的替代或补充方法[466-468]。许多检测杂交分子的电化学方法已经被提出，其中一部分正在用于商业化设备上。一些团队已经发表了基于大规模集成电路的传感器的文章，这些传感器利用电化学探测技术进行生物分子的片上探测[459,467,469]。

传感器设计。图 5.23 显示了智能 CMOS 图像传感器光和电化学双成像的显微照片。该传感器以 $0.35\mu m$(2 层多晶硅 4 层金属)的标准 CMOS 工艺制造。它包含一个由光学和电化学混合的像素阵列以及能够实现相关功能的控制/读出电路。这个混合像素阵列是 128×128 的光感应像素阵列，其中部分像素被电化学感应像素替代。光感应像素采用改进的 3T-APS，像素尺寸为 $7.5\mu m\times7.5\mu m$。电化学感应像素包含面积为 $30.5\mu m\times30.5\mu m$ 的裸露电极，同时用传输门开关进行行选，电化学感应像素的尺寸为 $60\mu m\times60\mu m$。这样，8×8 的光感应像素被电化学感应像素替代，该传感器有 8×8 个电化学像素阵列被嵌入在光图像传感器中。由于光图像传感器和电化学图像传感器在工作速度上存在很大的差异，光和电化学像素阵列被设计成独立工作模式。图 5.24 显示了该传感器的电路原理图。

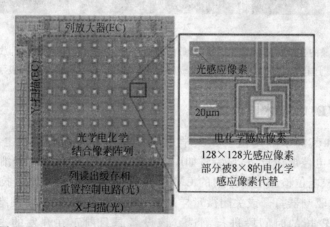

图 5.23　一个光和电化学双重成像智能 CMOS 传感器的显微照片

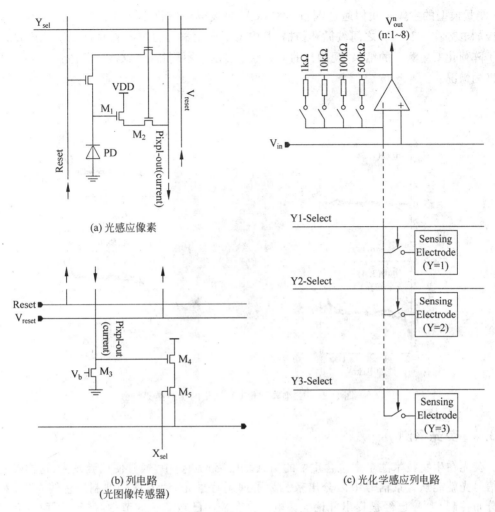

图 5.24 像素电路:(a)光感应;(b)电化学感应;(c)列级电路

压控电流测量方法被应用在片上双分子微阵列技术的电化学测量上。可选方法包括循环伏安法(cyclic voltammetry,CV)[468]和微分脉冲伏安法[466],这两种方法在发表的论文中称对 DNA 杂化探测是可行的。片上检测电流的电压跟随器可以进行多点的电化学测量。通过在电压跟随器(单位增益缓冲器)的反馈路径里加一个电阻,这个电路能实现压控电流的测量。这种电路结构已经广泛应用于电化学稳压器和膜片钳放大器。

实验结果。本实验进行二维阵列 CV 测量,并用一个单帧测量获得了 8×8 的 CV 曲线。对于片上测量,在电化学感应像素的铝电极上用金形成电极。由于金的化学稳定性和亲硫基性,它一直作为电化学分子测量的标准电极材料。Au/Cr(300nm/10nm)层被蒸镀到 $30.5\mu m\times 30.5\mu m$ 的电化学感应电极里,然后传感器被安装在陶瓷封装里并以铝线连接。用环氧橡胶层制成带有连接线的传感器的模具,其中只有混合像素阵列不被封装并在测量时裸露。

两电极结构被用在阵列化 CV 测量中,Ag/AgCl 电极作为对电极,工作电极是一个 8×8 阵列的金电极。这里采用一个在生理盐水中具有高电阻率的琼脂糖凝胶岛作为一个二维

CV 测量模型的主体。我们通过测量 8×8 CV 曲线得到电化学特性的图像。Ag/AgCl 对电极的电势在 −3～5V 之间被循环扫描,其中每个电化学行的扫描速度为 1Hz。图 5.25 显示了阵列化 CV 测量的结果,观测到的 CV 分布显示了不同特征,这些特征取决于各自测量电极的情况。

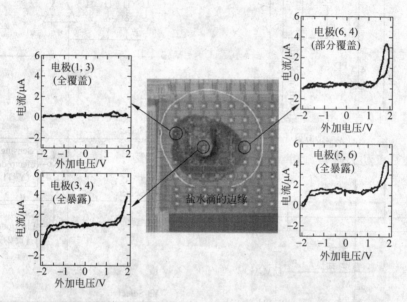

图 5.25 二维阵列化 CV 测量的实验结果

5.3.4 荧光探测

荧光在生物技术测量中包含重要的信息。在荧光测量中,通常使用激发光,这样荧光作为信号光就能从背景信号中区分出来。为了抑制背景光,片上信号抑制滤色镜[460-463]和电势分布控制[266,267]已经被提出并论证。电势分布控制已在 3.7.3 节中解释。下面将介绍一个用于小鼠脑内成像算法的智能 CMOS 图像传感器。

用于体内小鼠脑成像算法的智能 CMOS 图像传感器

智能 CMOS 图像传感器的一个重要应用是通过捕获大脑图像来研究它的学习和记忆功能[470]。目前的大脑成像技术需要昂贵的设备,该设备在图像精度和速度以及成像深度方面都有局限性,但这些性能对大脑研究都是很重要的[471]。如图 5.26 所示,一个小型化智能 CMOS 图像传感器(简称"CMOS 传感器")能够实时地以任意深度对大脑进行成像。

图像传感器实现。该图像传感器用标准 $0.35\mu m$ CMOS 工艺制造。如图 5.27 的芯片显微照片所示,它基于用脉冲宽度调制作为输出(PWM)的改进型 3T-APS。该传感器包含一个数字和模拟输出接口与外部读出电路,能够以接近视频速率输出图像的模拟信号。PWM 输出的数字输出方式适合长积分时间的静态成像。PWM 光传感器已在 3.4.1 节中进行了介绍。它被设计得足够大,可以给老鼠的海马体成像;还要足够小,可以给每个大脑半球进行创伤成像。它的规格列于表 5.4 中。值得一提的是,虽然它是可以集成有电刺激功能的传感器,但它具有唯一的成像功能。

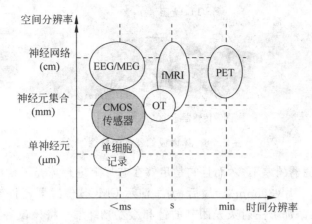

图 5.26　目前的神经元成像技术（EEG：脑电图；MEG：脑磁描图；OT：光学形貌图；fMRI：功能性磁共振成像；PET：正电子发射断层扫描）

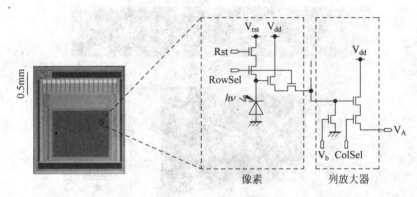

图 5.27　体内成像智能 CMOS 图像传感器的像素电路，右边的图显示了制造芯片的显微照片

表 5.4　体内成像智能 CMOS 图像传感器的规格

工艺	$0.35\mu m$ CMOS（2 层多晶硅 4 层金属）
像素数目	176×144（QCIF）
芯片尺寸	$2mm\times 2.2mm$
像素尺寸	$7.5\mu m\times 7.5\mu m$
光电二极管	n 阱/p 衬底结
填充因子	29%
像素类型	改进的 3T-APS

对于将传感器深植于老鼠大脑的应用来说，封装是一个关键问题。图 5.28 列举了一个封装器件，它集成了一个智能 CMOS 图像传感器和一个在柔性聚酰亚胺基板上用于激发的紫外 LED 灯。为了实现一个 $350\mu m$ 厚度的非常紧凑的器件，研究者开发了一套专用的制造工艺。

下面介绍制造工艺。该芯片首先被减薄到约 $200\mu m$，它连接到一个弹性的和生物相容的聚酰亚胺基板上并被一层透明的环氧树脂保护。将一种对于 7 氨基-4 甲基香豆素（AMC）的荧光发射具有高选择性的过滤层旋涂到器件的表面。通过这个简单的方法实现在 450nm 波长处的透过率为 $-44dB$，450nm 是荧光发射的波长。获得的透过率与用在荧光

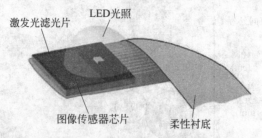

图 5.28 体内成像器件的概念图

显微镜中的离散过滤器的差不多,而且它是足够在芯片上进行荧光成像的。在传感器上附着紫外发射的 LED 芯片(365nm)。最后,为了演示 CMOS 传感器器件对大脑活动进行成像的操作,该器件又添加了注射荧光团基的针和激发光纤维。将该设备连接到 PC 来读出输出信号和输入控制信号。图 5.29 显示了开发完全的器件。

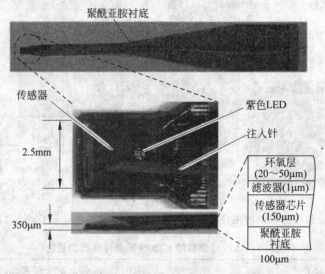

图 5.29 制造的体内成像器件的照片,包括传感器周围的特写

用这个器件连接荧光团基(PGR-MCA,VPR-MCA),实验检测老鼠海马体内丝氨酸蛋白酶(如纺锤菌素和组织型纤维蛋白溶酶原激活物)的存在。VPR-MCA 用来检测激活的纺锤菌素[472],同时 PGR-MCA 专门检测靶组织型纤维蛋白溶酶原激活物。在实验中,用红藻氨酸(KA)作为药剂,它可以导致在突触后神经元细胞外产生丝氨酸蛋白酶。带有插入器件的实验装置如图 5.30 所示。

体内成像的实验结果。实验中,可以通过 AMC 荧光成像实时观察丝氨酸蛋白酶的活性。由于丝氨酸蛋白酶的存在,AMC 荧光将从基板被释放。这里选择了多个在注射针出口附近不同位置的图像传感器并且将图像描绘出来,同一个位置的信号水平图像显示在图 5.31 中。从结果可以观察到,在 KA 注射后大约 1 小时荧光信号将有一个显著的增加。信号增加是注射针附近局部区域的丝氨酸蛋白酶活性增加的结果。为了证实这一观察结果,在实验结束时提取了老鼠大脑,在荧光显微镜下观察大脑切片。

本实验成功地证明了 CMOS 成像器件实时检测脑功能活动的能力。此外,通过使用体内丝氨酸蛋白酶的成像器件,实验独立地验证了 KA 对海马体的影响的发现。

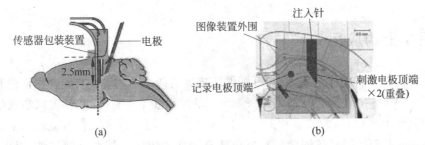

图 5.30 插入老鼠大脑的实验装置。(a)老鼠大脑内部器件的矢状面观;(b)老鼠大脑内部器件的冠状平面图

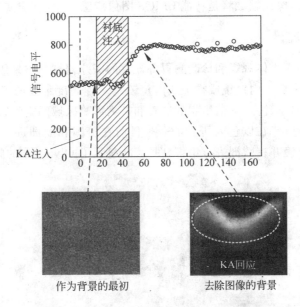

图 5.31 用智能 CMOS 图像传感器进行体内大脑深处成像的实验结果

另一项实验表明,在保证大脑继续工作并且能够正常地做出反应的条件下,用这种方法对大脑的损伤最小。

在接下来的步骤中,芯片上将集成激励电极。图 5.32 显示了一个带有紫外激发光源的体内智能 CMOS 图像传感器。为了紫外 LED 灯照明的使用,在芯片上制造了几个孔,这些孔也可用于 KA 的注射。

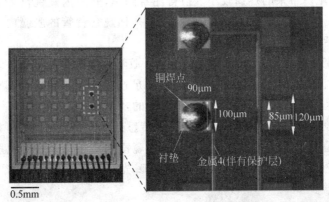

图 5.32 集成激励电极的高级体内成像智能 CMOS 图像传感器

5.4 医学上的应用

在这一部分中,提出了智能 CMOS 图像传感器的两个应用方向:胶囊内窥镜和视网膜假体。智能 CMOS 图像传感器适用于医学应用的原因如下。首先,它们可以集成信号处理器、射频和其他电子设备,即实现片上系统(SoC)是可能的。因为胶囊内窥镜要求系统体积小、功耗低,所以智能 CMOS 图像传感器很适合做这种应用。其次,对医疗应用来说,智能功能是很有用的。视网膜假体就是这样的一个例子,在视网膜假体芯片上添加一个电激励功能。在不久的将来,医疗领域将是智能 CMOS 图像传感器的最重要应用之一。

5.4.1 胶囊型内窥镜

内窥镜是一种插入人体观察和诊断肠胃等器官的医疗器械。它采用照亮被观察区域的导光纤维的 CCD 摄像机。内窥镜或推进式内窥镜是一种高度集成的仪器,它集成有相机、光导、摘取组织的小镊子、注水清洁组织的试管和扩大影响区域的空气管。胶囊内窥镜是由 Given Imaging 于 2000 年在以色列研发的[473],目前在美国和欧洲进行出售。Olympus 也研发了一种胶囊内窥镜并在欧洲进行销售,图 5.33 显示了 Olympus 的胶囊内窥镜。

图 5.33 胶囊内窥镜。胶囊长度为 26mm,直径为 11mm。胶囊主要部件包括圆顶盖、光学镜片、用于照明的白色 LED、CCD 摄像头、电池、RF 电路和天线(在此向 Olympus 致谢)

胶囊内窥镜使用一个图像传感器、光学成像系统、LED 照明灯、射频电路、天线、电池和其他组成部分。用户吞下胶囊内窥镜后,它会沿着消化器官自动移动。与传统的内窥镜相比,胶囊内窥镜给用户带来的疼痛较小。值得注意的是,胶囊内窥镜仅限于在小肠内使用,而不用于胃和大肠。最近,Given Imaging 研发了一种观察食管的胶囊内窥镜[474]。它用两个摄像头来对前后两侧的场景进行成像。

用于胶囊内窥镜的智能 CMOS 图像传感器。胶囊内窥镜是一种植入装置,因此尺寸和功耗是关键问题,智能 CMOS 图像传感器能够满足这些要求。在应用 CMOS 图像传感器时,必须考虑彩色实现方法。正如在 2.11 节讨论的,CMOS 图像传感器采用滚筒式曝光机制。医疗用途一般需要色彩重现,所以三个图像传感器或三个光源是更好的选择。事实上,传统的内窥镜正是使用了三光源的方法。胶囊内窥镜需要在一个小体积里安装摄像系统,因此三光源的方法特别适用。然而,由于滚筒曝光机制,三个光源的方法不能应用于 CMOS 图像传感器。

在滚筒曝光中,每行的快门时间是不同的,因此每一个光源在不同时间发光的三光源方法不能应用。目前销售的胶囊内窥镜在 CMOS 图像传感器(Given Imaging)或者 CCD 图像传感器(Olympus)上集成芯片级滤光片。为了在 CMOS 图像传感器里应用三光源的方法,需要一个全局快门。有研究者提出了另一种计算实现色彩再现的方法。当应用三光源方法时,由于在滚动快门中 RGB 混合比是已知的,因此可以在芯片外单独计算 RGB 分量[475]。虽然颜色再现能力必须详细评估,但对使用三光源方法的 CMOS 图像传感器而言,这是一种很有前途的方法。

由于胶囊内窥镜用电池工作,整个电子设备必须做到低功耗,同时,总体积也应该较小,因此,包括射频电路和成像系统的 SoC 是很好的选择。关于采用 CMOS 图像传感器和射频电路的 SoC 已经有相关文章发表[476]。如图 5.34 所示,除了集成了二进制相移键控(BPSK)调制电子的电源(VDD 和 GND)外,所制造的芯片只有一个数字 I/O 端口。在 QVGA 格式 2fps 条件下,该芯片功耗为 2.6mW。已经有人发表了用于胶囊内窥镜的 SoC 的文章,但是图像传感器是分离的,并没有集成在 SoC 中。该系统具有以 2Mbps 的速度无线传输 320×288 个像素数据的能力且功耗为 6.2mW[477]。

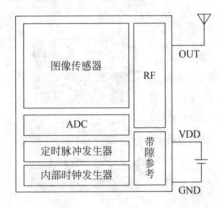

图 5.34 胶囊内窥镜中含有 CMOS 传感器的 SoC,本芯片使用环状 ADC[476]

胶囊内窥镜的另一个令人满意的功能是片上图像压缩。现在已经有几篇关于片上压缩的文章[478-481],预计在不久的将来这个功能将被应用在胶囊内窥镜上。这些 SoC 将结合其他技术并应用于胶囊内窥镜中,例如微电子机械系统(MEMS)、微全分析系统(μTAS,LOB)来监测其他物理量(如电势、pH 值和温度等)[482,483]。如 5.3.1 节所述,这种多模态感应适合智能 CMOS 图像传感器。

5.4.2 视网膜假体

在该领域的早期工作中,MOS 图像传感器已经被用来帮助盲人。Optacon,又叫盲人阅读器,可能是第一个用于帮助盲人的固态图像传感器[484]。Optacon 集成了扫描电路和读出电路,并且拥有很小的体积[119,485]。视网膜假体是植入式的 Optacon。在 Optacon 中,盲人通过触觉看到物体,而在视网膜假体中,盲人通过植入的装置给予视觉相关细胞电激励来感知物体。

该领域已经进行了大量的研究,针对不同的植入部位[486],如皮质区[487,488]、视神经[489]、视网膜外的空间[490]和视网膜下的空间[497-503]。最近另一种被称为脉络膜上腔反式视黄醛激励(STS)的方法已经被提出来[504-506]。

视网膜的空间植入或眼植入能防止感染,适用于除感光细胞之外的视网膜细胞仍然能够正常工作色素性视网膜炎(RP)患者和(与年龄相关的)老年性黄斑变性(AMD)患者。值得注意的是,RP 和 AMD 还没有有效的治疗方法。视网膜的结构示意图为附录 C 的图 C.1。

虽然在外视网膜方法中,是给神经节细胞激励,但是在亚视网膜方法中,激励仅仅替代了感光细胞,这样,在这个方法的实现上,双极细胞和神经节细胞有可能同时被激励。因此,亚视网膜的方法与外视网膜的方法相比具有以下的优点:激励点和视觉感应能够很好符合,还可以自然地利用光学功能,如眼球运动和虹膜的开合。图 5.35 显示了三种方法:外视网膜激励、亚视网膜激励和 STS。

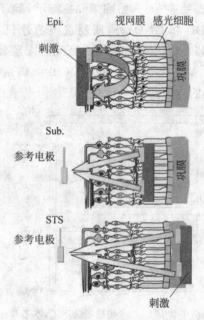

图 5.35　用于眼睛移植的视网膜假体的三种方法

使用 PFM 光传感器的亚视网膜方法。在亚视网膜植入中,为了集成激励电极还需要一个光传感器。目前采用的光敏像素阵列是简单的无任何偏置电压(即太阳能电池模式下)的光电二极管阵列,这主要是因为其结构简单并且不需要电源[497-499]。光电流直接用作进入视网膜细胞的激励电流。在日光环境下,为了生成足够的激励电流,用于亚视网膜的脉冲频率调制(PFM)光传感器已经被提出了[201,212],同时基于 PFM 的视网膜假体器件和用于STS 的模拟器也已经被研发出来了[190,193,202-211,507-511]。近年来,几个团队也已经研发了用于亚视网膜下植入的 PFM 光传感器或基于脉冲的光传感器[213-216,512-515]。视网膜假体眼睛移植的整体系统如图 5.36 所示。

PFM 适合作为在视网膜下植入的人工视网膜装置,理由如下。首先,PFM 产生脉冲流输出,这很适合用来激励细胞。一般情况下,对于刺激细胞的电位来说脉冲激励非常有效。此外,这样的脉冲形式是和逻辑电路兼容的,这样就能实现多样化的功能。第二,PFM 可以在不降低信噪比的情况下,在非常低的供电电压下工作,这很适合于植入式装置。最后,它的光灵敏度足够高,可以在正常照明条件下进行检测并且它的动态范围很大。这些特点决定 PFM 像素很适合替代感光细胞。虽然基本上 PFM 光传感器对人工视网膜装置的应用很适合,但是还有一些待改进的地方,本文将对这些内容进行描述。

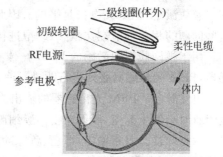

使用图像传感器的外视网膜方法。值得一提的是,当用激励成像时,亚视网膜的方法让人感觉更加自然,因为成像可以在与正常眼球相同的激励平面内完成。然而,一些外视网膜的方法,也可以使用带有激励的成像装置。如 3.6.1.2 节所述,蓝宝石上硅(SOS)是透明的,它可以在应用背照式图像传感器时作为外视网膜。对于背照式结

图 5.36 视网膜假体眼睛移植的整体系统

构,成像区和激励可以放置在同一平面。使用 SOS CMOS 技术的 PFM 传感器已被证明可以用于外视网膜[203]。

另一个采用图像传感器的外视网膜方法是利用三维(3D)集成技术[213,512,513]。图 5.37 显示了采用三维集成技术的视网膜假体的概念。

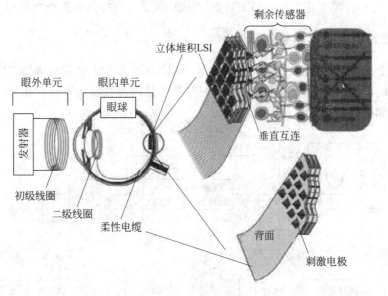

图 5.37 用三维集成技术的外视网膜方法[213](在此感谢东北大学的 M. Koyanagi 教授)

眼睛植入的大规模集成电路(LSI)。应该指出的是,当将基于 LSI 的仿真器件应用到视网膜假体上时需要克服许多技术上的问题。虽然许多外视网膜的方法已经使用了 LSI[492,494,516],但亚视网膜方法在使用 LSI 时存在困难,因为它是完全植入组织的,所以必须同时为图像传感器和电仿真器件工作。

生物相容性。基于 LSI 的接口必须具有生物相容性。标准 LSI 结构不适用于生物环境,通常情况下氮化硅作为标准 LSI 的保护层,但长期在生物环境中会有损坏。

与标准 LSI 兼容的制造。激励电极必须与标准 LSI 结构兼容。在标准 LSI 中,用作输入输出接口的焊盘是由铝制成的,但是由于铝在生物环境下具有溶解性,所以用其作为视网膜细胞的激励电极是行不通的。铂金是当前做激励电极的一种候选材料。

激励电极的形状。除了电极材料外,激励电极的形状也影响激励的效率。凸形更容易获得有效激励,但 LSI 电极是平的。因此,在 LSI 中如何制作凸形铂激励电极是一项挑战。这些问题的相关细节在文献[212]中进行了讨论。

用于视网膜细胞激励的 PFM 光传感器的改进。为了在视网膜细胞中应用 PFM 光传感器,PFM 光传感器必须进行改进。改进的理由如下。

电流脉冲。PFM 光传感器的输出为电压脉冲波形的形式,然而电极和细胞之间的接触电阻有时会发生改变,所以向视网膜细胞中注入电荷最好是电流输出。

双相脉冲。双相输出,即正、负脉冲,可以在视网膜细胞的电激励下保持电荷平衡。在临床使用中,电荷平衡是一个关键的问题,因为在活体组织中残留电荷的积累可能对视网膜细胞产生有害的影响。

频率限制。输出频率的限制是因为过高的频率可能会引起视网膜细胞的损害。如图 3.11 所示,对激励视网膜细胞来说传统的 PFM 器件的输出脉冲频率(约 1MHz)过高,但是频率限制会引起输入光强范围的减小。为了缓解这个问题,需要引入可变灵敏度机制,其中输出频率用分频器分为 2^{-n} 份。这个想法的灵感来自于动物视网膜的光适应机制,如附录 C 图 C.2 所示。需要注意的是,PFM 数字输出和引入的分频器逻辑功能相兼容的。

PFM 光传感器视网膜模拟器

基于以上改进,像素电路采用标准的 $0.6\mu m$ CMOS 像素工艺设计和制造[202]。图 5.38 显示了像素的框图。频率限制通过使用开关电容的低通滤波器实现。双相脉冲电流由交替切换电流源和接收点来实现。

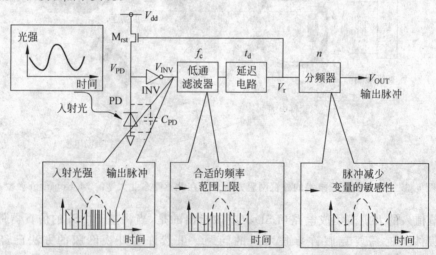

图 5.38 用于视网膜细胞激励的改进 PFM 光传感器像素电路框图

图 5.39 显示了使用可变光灵敏度芯片的实验结果。原输出曲线的动态范围超过 6-log(输入光强的第六阶范围),但当低通滤波器打开时,频率被限制在 250Hz,动态范围将减少到大约 2-log。通过引入可变灵敏度,输入光强总范围变为 $n=0$ 和 $n=7$ 之间的 5-log,其中是 n 是灵敏度分段数。

使用 PFM 光传感器的图像处理。当 PFM 型模拟器装置应用于视网膜假体时,分辨率通常小于 30×30。这种限制是因为电生理实验电极间距大于 $100\mu m$,并且植入的芯片的宽

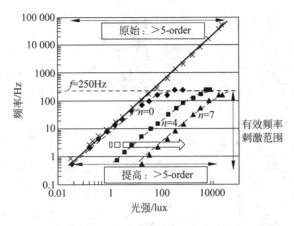

图 5.39 可变光灵敏度的 PFM 光传感器的实验结果

度小于 3mm。为了在低分辨率下获得清晰的图像,需要边缘增强之类的图像处理操作。为了实现图像处理,研究人员进行了在脉冲频域实现空间滤波的相关研究[190],描述如下。

空间滤波通常是基于使用内核 h 的空间相关操作

$$g(x,y) = \sum_{x'}\sum_{y'}h(x',y')f(f+x',y|+y') \qquad (5.1)$$

其中 $f(x,y)$ 和 $g(x,y)$ 分别表示输入和输出图像在点 (x,y) 处的像素值。

通常 f、g、h 是模拟图像处理的模拟值或数字图像处理的整数值。在这个方案中,f 和 g 表示脉冲频率。因此,为了实现图像处理功能,研究了一种代表脉冲域的内核权重方法。于是引入了与相邻像素的相互作用,这种相互作用表现为对相邻像素的脉冲流的栅控制。

这个概念如图 5.40 所示。内核权重的绝对值 $|h|$ 表示栅控制的开关频率。为了在空间滤波内核中实现负权重,来自一个像素的脉冲与来自相邻像素的脉冲相互作用来使它们消失。对于正的权重,像素脉冲和来自相邻像素的脉冲合并。这些机制可以利用简单的数字电路来实现。在该架构中,用 1bit 脉冲缓冲存储器来吸收相互作用脉冲之间的相位失配。值得注意的是,这里的操作存在随机过程特性[191],因此可以替换为另一个结构,如采用脉冲编码的生物信息处理[188]。

所提出的架构可以执行基本的图像处理,如边缘增强、模糊和边缘检测。与简单数字空间滤波相比,它不需要加法器和乘法器,因此只需要少量的逻辑门。

二值化电路是由一个 $n$bit 的 D 触发器(D-FF)组成的异步计数器实现的。D-FF 的 MSB 的输入固定在高电平。当 $2N-1$ 个脉冲输入时,计数器的输出由高变低。这意味着这个计数器是一个以 $2N-1$ 为固定阈值的数字比较器。

根据这种结构,一个拥有 16×16 像素的基于 PFM 的视网膜假体得到了验证。图 5.41 显示了所制造芯片的外围电路和像素的显微照片。该芯片实现了两个相邻(上面(北)和左面(西))像素的相关性。如图 5.41 所示,每个像素有一个 PFM 光传感器、一个图像处理电路、一个激励电路和一个激励电极,也就是说该芯片可以激励视网膜细胞。该芯片用于在体外电生理实验中刺激视网膜细胞,下一节将对此进行详细描述。图 5.42 显示了使用该芯片进行图像处理的实验结果,图中包含原始图像、边缘增强后和模糊后的图像。这些结果清楚地表明,所提出的架构能很好地工作。

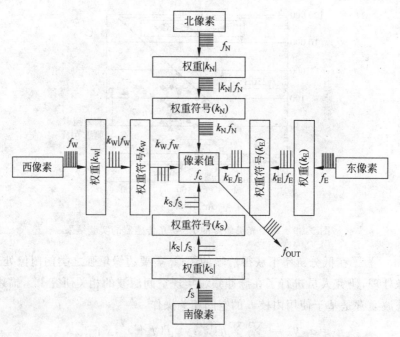

图 5.40 频域图像处理的概念图

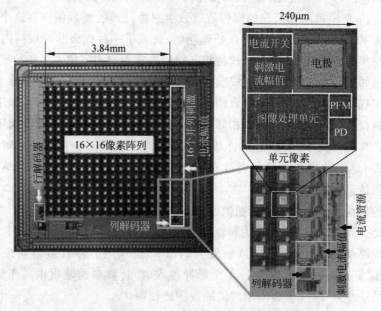

图 5.41 制造的视网膜假体 PFM 芯片的显微照片

接下来验证像素和相关像素数目进一步增加的 PFM 光传感器芯片的实验结果。该芯片有 32×32 像素,其中电容反馈 PFM 用 4 个相邻像素的相关处理来实现,如图 5.43 所示,但是该芯片无视网膜激励电路[209]。电容反馈 PFM 在 3.4.2.2 节进行了描述。实验结果如图 5.44 所示,获得了平滑图像操作的预处理结果。

PFM 光传感器在视网膜细胞激励中的应用。本节将证实上一节所介绍的基于 PFM 的仿真单元在激励视网膜细胞时是有效的。为了将 Si-LSI 芯片应用到电生理学实验中,必须保护这些芯片免受生物环境的影响,并且制作一个与标准 LSI 结构相匹配的有效激励电极。为了满足这些要求,研究人员研发出了 Pt/Au 堆叠的凸起电极[508-511]。然而,限于本书篇幅,这里不对这个电极进行讨论。

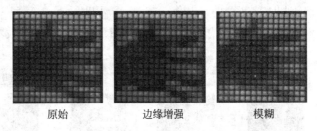

图 5.42 使用两个相邻像素相关的 PFM 图像传感器的图像处理结果[190]

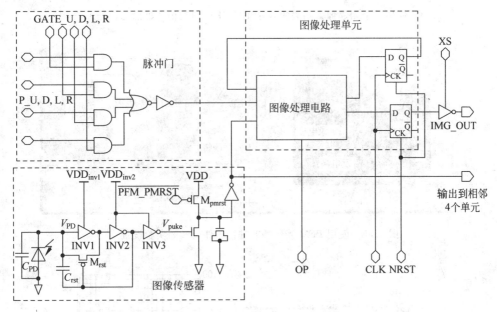

图 5.43 具有图像处理功能的电容反馈 PFM 的像素电路图[209]

为了验证 PFM 光传感器芯片的工作情况,使用分离的牛蛙视网膜进行体外实验。本实验中,芯片作为一种通过输入光强控制的激励,它只起到感光细胞的作用。芯片上集成了一个电流源和脉冲整形电路。文献[508-511]还完成了 Pt/Au 堆叠的凸起电极和芯片成型工艺。

一片牛蛙视网膜和侧脸上的视网膜神经节细胞(RGC)被放置在封装芯片的表面上。图 5.45 显示了实验装置。电刺激是在芯片上的某个选定的像素上进行的。视网膜上放置了一个末端直径 5μm 的钨计数电极,将在计数电极与芯片电极之间产生一个视网膜传导电流。实验中使用阴极优先的双相电流脉冲作为激励。该脉冲的相关参数请参考图 5.45 的插图中的描述。详细的实验装置在参考文献[208]中给出。值得注意的是,近红外光不刺激视网膜细胞,但会激发 PFM 光传感器细胞。图 5.46 还显示了用 PFM 光传感器刺激视网膜细胞的实验结果,其中通过输入近红外光进行照明。发射率随输入近红外光强成比例增加。这表明 PFM 光传感器可以通过近红外光输入来激活视网膜细胞,同时表明它可以应用于人类视网膜假体。

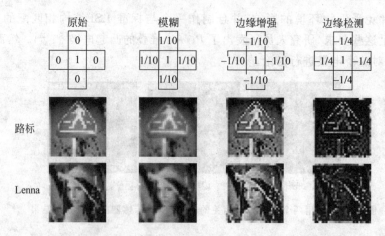

图 5.44 使用四个相邻像素相关的电容反馈 PFM 图像传感器的图像处理结果[209]

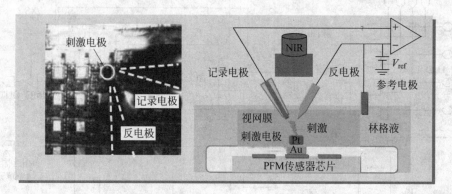

图 5.45 使用 PFM 光传感器进行体外刺激的实验设置[208]

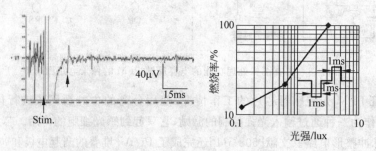

图 5.46 用 PFM 光传感器进行体外激励的实验结果：(a)得到的示例波形；(b)发射率是输入光强度的函数[208]

附录A 常量表

表 A.1 300K下物理常数

常量	符号	数值	单位
阿伏伽德罗常数	N_{AVO}	6.02204×10^{23}	mol^{-1}
玻尔兹曼常数	k_B	1.380658×10^{-23}	J/K
电子电荷	e	$1.60217733 \times 10^{-16}$	C
电子质量	m_e	$9.1093897 \times 10^{-31}$	kg
电子电压	eV	$1eV = 1.60217733 \times 10^{-16} J$	
真空中磁导率	μ_0	1.25663×10^{-6}	H/m
真空中介电常数	ε_0	$8.854187817 \times 10^{-12}$	F/m
普朗克常数	h	$6.6260755 \times 10^{-34}$	J·s
真空中光速	c	2.99792458×10^8	m/s
300K下热电压	$k_B T$	26	meV
1fF电容热噪声	$\sqrt{k_B T/C}$	5	μV
1eV量子波长	λ	1.23977	μm

表 A.2 300K下材料的特性

属性	单位	硅	锗	二氧化硅	氮化硅
禁带宽度	eV	1.1242	0.664	9	5
介电常数		11.9	16	3.9	7.5
折射率		3.44	3.97	1.46	2.05
本征载流子浓度	cm^{-3}	1.45×10^{10}	2.4×10^{13}	—	—
电子迁移率	cm^2/Vs	1430	3600	—	—
空穴迁移率	cm^2/Vs	470	1800	—	—

附录B 光 照 度

图B.1显示了不同光线情况下的典型光照度级别。人类视觉的最低阈值大约是 10^{-6} lux[390]。

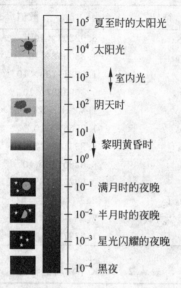

图 B.1　不同光线情况下的典型光照度级别[6,143]

辐射度与光照度的关系。表B.1为辐射度与光照度的计量汇总，图B.2为适光眼的响应。适光量到物理量的转换因数 K 为

$$K = 683 \frac{\int R(\lambda) V(\lambda) d\lambda}{\int R(\lambda) d\lambda} \tag{B.1}$$

典型转换因数汇总在表 B.2[143]。

表 B.1　辐射量与光照量 $K(\mathrm{lm/W})$[143]

辐射量	辐射单位	光照量	光照单位
辐射强度	W/sr*	光照强度	candela(cd)
辐射通量	W=J/S	光通量	lumen(lm)=cd·sr
辐照度	W/m²	光照度	lm/m²=lux
辐射率	W/m²/sr	亮度	cd/m²

* sr：立体角(steradian)。

成像平面的光照度。在这部分中我们介绍传感器成像平面上的光照度[6]。lux 为常见的光照度单位。值得注意的是，lux 是一个有关人眼特性的光度测定单位，也就是说，它不是纯粹的物理单位，光照度被定义为单位面积上的光功率。假定一个光学系统如图 B.3 所示。光通量 F 是照射到一个表面可以视为完全扩散的物体上。当光被一个反射率为 R、面

积为 A 的完美扩散表面的物体反射时,反射光均匀分散在整个立体角 π 的一半区域上。因此光通量 F_o 在一个单位立体角上的分布公式为

$$F_o = \frac{FR}{\pi} \tag{B.2}$$

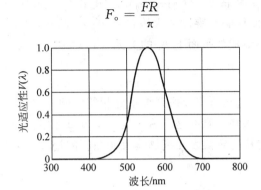

图 B.2　适光眼响应

表 B.2　典型转换因数[143]

光源	转换因数 $K/(lm/W)$
555nm 绿光	683
红色 LED	60
没有云的日光	140
2850K 标准光源	16
带有红外滤波器的 2850K 标准光源	350

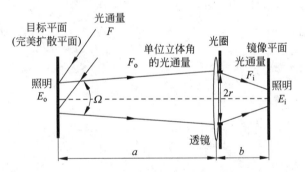

图 B.3　在成像平面上的光照度

把立体角变换到光圈或虹膜 Ω 上为

$$\Omega = \frac{\pi r^2}{a^2} \tag{B.3}$$

通过透射率为 T 的透镜进入传感器成像平面的光通量 F_i 的公式为

$$F_i = F_o \Omega = FRT \left(\frac{r}{a}\right)^2 = FRT \frac{1}{4F_N^2}\left(\frac{m}{1+m}\right)^2 \tag{B.4}$$

假定透镜放大倍数为 $m=b/a$,焦距为 f,光圈 F 为 $F_N=f/(2r)$,并且

$$a = \frac{m}{1+m}f \tag{B.5}$$

因此

$$\left(\frac{r}{a}\right)^2 = \left(\frac{f}{2F_N}\right)^2 \left[\frac{m}{(1+m)f}\right]^2 \tag{B.6}$$

光照度在物体和传感器成像平面上分别为 $E_o = F/A$ 和 $E_i = F_i/(m^2 A)$。值得注意的是,物体聚焦到成像平面上的面积需要乘以放大因数 m 的平方。通过上述公式我们可以得到光照度在物体上 E_o 和传感器成像平面上 E_i 的关系式:

$$E_i = \frac{E_o RT}{4F_N^2(1+m)^2} \approx \frac{E_o RT}{4F_N^2} \tag{B.7}$$

其中第二个等式中 $m \ll 1$,这与传统成像系统正好相符。例如,当 F_N 为 2.8 且 $T=R=1$ 时,E_i/E_o 正好大约为 1/30。值得注意的是 T 和 R 通常小于 1,因此这个比例通常小于 1/30。另外还要注意的是,光照度从传感器表面到物体上光照度的过程中衰减 1/10~1/100。

附录 C 人眼和 CMOS 图像传感器

在这部分中,我们将介绍人眼的视觉过程,因为它们是理想的成像系统和 CMOS 成像系统的模型。在最后,我们将比较人类视觉系统和 CMOS 图像传感器。

视网膜。即使和世界上最先进的 CMOS 图像传感器相比,人眼也拥有更为优越的特性。人眼的动态范围大约为 200dB,且为多分辨率。另外人眼有时空图像预处理的焦平面处理功能,并且人类拥有两只眼睛,这使得我们可以通过聚焦和视差进行测距。值得注意的是,利用视差测量距离需要在视觉皮层进行一系列复杂的处理[517]。人类的视网膜面积约为 5cm×5cm,厚度约为 0.4mm[187,300,392,518]。其概念性结构如图 C.1 所示。入射光是由视杆细胞和视锥细胞探测的。

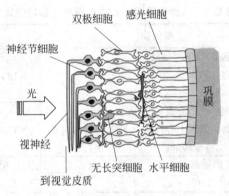

图 C.1 人类视网膜结构[300]

人眼的感光性。如图 C.2 所示,视杆细胞与视锥细胞相比,具有更高的感光灵敏度并且对光强有自适应性。在均匀的光照环境中,视杆细胞工作在带有饱和特性的两个数量级的范围内。图 C.2 以图解的方式显示了在恒定光照下的光响应曲线。光响应曲线根据环境光照进行自适应移位,最终变换了七个数量级。人眼由于其自身的机理在月光和日光下都拥有大动态范围。

视网膜上的颜色。人眼可以感知波长为 370~730nm 范围内光的颜色。视杆细胞主要分布在视网膜的外围,虽然不能感知颜色,但它有很高的光敏性,而视锥细胞主要集中在视网膜或凹点中心,它可以感知颜色,但比视杆细胞的感光性要差,因此视网膜具有高、低感光度的两种感光结构。通常在暗光条件下,主要是视杆细胞感光,这种视觉模式被称为暗视觉;而当由视锥细胞感知光时,被称为明视觉。暗视觉和明视觉的峰值波长分别为 507nm 和 555nm。对于颜色敏感度,视杆细胞被分为 L、M 和 S 型,这分别与图像传感器芯片上的 R、G 和 B 滤波器对应。L、M 和 S 型视锥细胞的中心波长分别为 565nm、545nm 和 440nm。不同的动物具有不同色感度,例如有些蝴蝶可以感知紫外线范围的光。令人惊讶的是,L、M 和 S 型视锥细胞的分布并不是均匀的,而图像传感器的颜色滤波器则是规律的排列,例如贝尔模板。

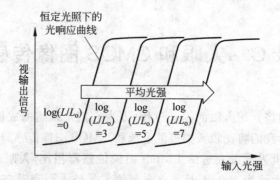

图 C.2 视杆细胞对光强的视觉调整。光感应曲线根据平均环境照度 L 改变,从初始值 L_0 开始,环境照度在 $\log(L/L_0)=0\sim7$ 的范围内以指数方式变换

人类视网膜与 CMOS 图像传感器的比较。表 C.1 汇总了人类视网膜与 CMOS 图像传感器的比较[187,300,392,518]。

表 C.1 人类视网膜与 CMOS 图像传感器的比较

项目	视网膜	图像传感器
分辨率	视锥细胞 5×10^6 视杆细胞 10^8 神经节细胞 10^6	$1\sim10\times10^6$
尺寸	视杆细胞:近凹心直径 $1\mu m$ 视锥细胞:黄斑中心凹内直径 $1\sim4\mu m$,外部 $4\sim10\mu m$	$2\sim10\mu m$ sq.
颜色	3 种视锥细胞(L,M,S) (L+M):S=14:1	片上 RGB 滤波器 R:G:B=1:2:1
最小探测光强	~0.001lux	$0.1\sim1$lux
动态范围	超过 140dB(自适应)	$60\sim70$dB
探测方法	顺反异构化→两级放大 (500×500)	电子空穴对产生 电荷收集
响应时间	~10ms	帧率(视频速率:33ms)
输出	脉冲频率调制	模拟或数字电压
输出数量	GCs 数:~1M	一个模拟输出或者 bit 数的数字输出
功能	光电转换 自动适应功能 时空信号处理	光电转换 放大 扫描

附录 D MOS 电容的基本特性

MOS 电容是由一个金属电极（通常用重掺杂多晶硅）和半导体之间加一个绝缘体（通常是 SiO_2）组成的。MOS 电容是 MOSFET 的重要组成部分，在标准 CMOS 工艺中通过连接 MOSFET 的源极和漏极很容易实现 MOS 电容。在本书中，MOSFET 的栅极和衬底作为电容的电极。MOS 电容的特性主要由绝缘体或者 SiO_2 下面的沟道决定。需要注意的是，MOS 电容是由两个电容串联构成的：一个栅氧化层电容 C_{ox} 和一个耗尽区电容 C_D。

MOS 电容有三种状态：积累、耗尽和反型，如图 D.1 所示，它是通过表面电势 Ψ_s 表征的。表面电势 $e\Psi_s$ 被定义为表面（$z=0$）和体硅区（$z=\infty$）之间能隙的能量差。

$\Psi_s < 0$：积累模式。

$\Psi_B > \Psi_s > 0$：耗尽模式。

$\Psi_s > \Psi_B$：反型模式。

这里的 $e\Psi_B$ 被定义为费米能级 E_{fs} 和体硅区 $E_i(\infty)$ 的中间能隙之间的能量差。在积累模式中，栅极偏压是负的，$V_g < 0$，空穴积累在表面附近，这种模式很少用在图像传感器中。在耗尽模式中，栅极施加正偏压 $V_g > 0$，使表面区域的自由载流子耗尽。电离后的受主空间电荷位于耗尽区，通过栅极偏压 V_g 补偿感应电荷。在这种模式中，表面势垒 Ψ_s 是正的但小于 Ψ_B。第三个模式是反型模式，它是用在 MOSFET 导通状态，并且在 CMOS 图像传感器中积累光生电荷。在耗尽模式中施加更大的栅极偏压时，将会出现反型层，其中电子在表面区域内积累。当 E_i 在 $z=0$ 处与 E_{fs} 相交，则进入反型模式，其中 $\Psi_s = \Psi_B$。假如 $psi_s > 2\Psi_B$，表

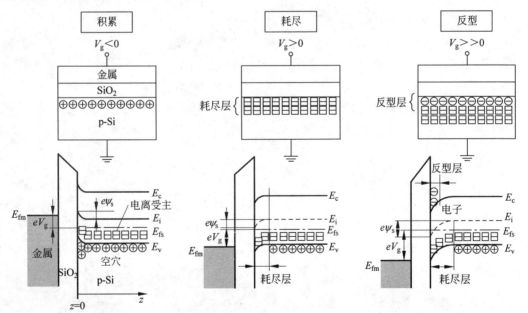

图 D.1 MOS 电容工作的三种状态的概念插图，积累、耗尽和反型。E_c、E_v、E_{fs} 和 E_i 分别是导带底、价带顶和半导体的费米能级以及中间能级。E_{fm} 是金属的费米能级，V_g 是栅极偏压

面将完全反型,即成为了 N 型区,这种模式被称为强反型;当 $\Psi_s < 2\Psi_B$ 时,称为弱反型。需要注意的是,反型层中的电子是热运动产生的电子或者是扩散电子,因此需要一定的时间来建立电子反型层,即非平衡状态下的反型层可以作为电子(例如入射光产生的)存储器。这个存储器被称为存储光生载流子的势阱,如果源极和漏极位于反型层的任意一侧。例如在 MOSFET 中,电子迅速地从源极和漏极中输送到反型层中,可以在很短的时间内建立一个充满电子的反型层。

附录 E MOSFET 的基本特性

增强型和耗尽型

MOSFET 被分为两种类型：增强型和耗尽型。尽管在一些传感器中使用耗尽型 MOSFET，但是，通常情况下都是使用增强型 CMOS 传感器。增强型 NMOSFET 的阈值电压为正值，而耗尽型 NMOSFET 的阈值电压为负值。所以耗尽型 NMOS 场效应管可以在没有施加栅极电压时开启，即处于常开状态。在像素电路中，阈值电压是非常重要的，所以，像素中的一些传感器中使用耗尽型 MOSFET[116]。

工作区：

MOSFET 的操作首先分为两个区域：高于阈值和低于阈值（亚阈值）。在每个区域，存在三个子区域，即：截止区、线性区和饱和区。在截止区没有漏极电流流过。这里，我们总结了 NMOS 在每个区的特点。

高于阈值：$V_{gs} > V_{th}$。

线性区：高于阈值的线性区的条件是

$$V_{gs} > V_{th}$$
$$V_{ds} < V_{gs} - V_{th} \tag{E.1}$$

在上述条件下，漏极电流 I_d 被表示为

$$I_d = \mu_n C_{ox} \frac{W_g}{L_g} \left[(V_{gs} - V_{th}) V_{ds} - \frac{1}{2} V_{ds}^2 \right] \tag{E.2}$$

其中 C_{ox} 是栅氧化层的单位面积的电容，W_g 和 L_g 是栅极宽度和长度。

饱和区：

$$V_{gs} > V_{th}$$
$$V_{ds} > V_{gs} - V_{th} \tag{E.3}$$

上述条件下

$$I_d = \frac{1}{2} \mu_n C_{ox} \frac{W_g}{L_g} (V_{gs} - V_{th})^2 \tag{E.4}$$

对于短沟道晶体管，必须考虑沟道长度调制效应，因而式(E.4)修改为[126]

$$I_d = \frac{1}{2} \mu_n C_{ox} \frac{W_g}{L_g} (V_{gs} - V_{th})^2 (1 + \lambda V_{ds}) = I_{sat}(V_{gs})(1 + \lambda V_{ds}) \tag{E.5}$$

其中 $I_{sat}(V_{gs})$ 是在没有沟道长度调制、栅源电压为 V_{gs} 时的饱和漏电流。式(E.5)意味着，即使在饱和区，漏极电流也会随着漏-源电压逐渐增加而增加。在双极晶体管中，一个类似的效应被称为厄利效应，特征参数是厄利电压 V_E。根据式(E.5)，在 MOSFET 中的厄利电压 V_E 为 $1/\lambda$。

亚阈值区：

在该区域中，满足下列条件

$$0 < V_{\text{gs}} < V_{\text{th}} \tag{E.6}$$

在这种情况下,仍然具有漏极电流,并表示为[523]

$$I_d = I_0 \exp\left[\frac{e}{mk_BT}\left(V_{\text{gs}} - V_{\text{th}} - \frac{mk_BT}{e}\right)\right]\left[1 - \exp\left(-\frac{e}{k_BT}V_{\text{ds}}\right)\right] \tag{E.7}$$

这里 m 是体效应系数[524],以后再定义,I_0 由下式给出:

$$I_0 = \mu_n C_{\text{ox}} \frac{W_g}{L_g} \frac{1}{m}\left(\frac{mk_BT}{e}\right)^2 \tag{E.8}$$

直观的提取上述亚阈值电流的方法见文献[42,171]。一些智能图像传感器利用亚阈值工作,因此我们在文献[42,171]处理后简要地考虑亚阈值电流的来源。

在亚阈值区的漏极电流源于扩散电流,这是由于源端和漏端之间的电子密度 n_s 和 n_d 之差,即

$$I_d = -qW_g x_c D_n \frac{n_d - n_s}{L_g} \tag{E.9}$$

其中 x_c 是沟道深度。值得注意的是,每端的电子密度 n_s 和 n_d 是由电子的势垒高度 $\Delta E_{s,d}$ 决定的,由此

$$n_{s,d} = n_0 \exp\left(-\frac{\Delta E_{s,d}}{k_BT}\right) \tag{E.10}$$

这里 n_0 为常数,每端的势垒能量由下式给出

$$\Delta E_{s,d} = -e\Psi_s + e(V_{\text{bi}} + V_{s,d}) \tag{E.11}$$

其中 Ψ_s 是在栅极的表面电位。在亚阈值区的,Ψ_s 大致是栅极电压 V_{gs} 的线性函数

$$\Psi_s = \Psi_0 + \frac{V_{\text{gs}}}{m} \tag{E.12}$$

这里 m 是由下式给定的体效应系数

$$m = 1 + \frac{C_d}{C_{\text{ox}}} \tag{E.13}$$

其中 C_d 为单位面积的耗尽层的电容。值得注意的是,$1/m$ 为测量从栅极到沟道电容耦合率。利用式(E.9)、式(E.10)、式(E.11)、式(E.12),当源电压接地时可以得到式(E.7)。亚阈值斜率 S 通常用于测量的亚阈特性,并定义为

$$S = \left[\frac{d(\log_{10} I_{\text{ds}})}{dV_{\text{gs}}}\right]^{-1} = 2.3\frac{mk_BT}{e} = 2.3\frac{k_BT}{e}\left(1 + \frac{C_d}{C_{\text{ox}}}\right) \tag{E.14}$$

S 的值通常是 $70\sim100\text{mV/decade}$。线性区域的亚阈值区的也分为线性区、饱和区,以及高于阈值的区域。在线性区中,I_d 依赖于 V_{ds};而在饱和区,I_d 几乎是独立于 V_{ds} 的。

在线性区中,V_{ds} 很小而且是由漏极和源极流出的扩散电流,根据条件 $V_{\text{ds}} < \frac{k_BT}{e}$,可以得到

$$I_d = I_0 \exp\left[\frac{e}{mk_BT}(V_{\text{gs}} - V_{\text{th}})\right]\frac{e}{k_BT}V_{\text{ds}} \tag{E.15}$$

这说明,I_d 是 V_{ds} 的线性表示。

饱和区：

在饱和区，V_{ds} 比 $\dfrac{k_B T}{e}$ 值大，因而式(E.7)变为

$$I_d = I_0 \exp\left[\frac{e}{m k_B T}(V_{gs} - V_{th})\right] \tag{E.16}$$

在这个区域内，漏极电流是独立于漏源电压的，并且当源电压恒定时仅依赖于栅极电压。从线性区过渡到饱和区附近发生 $V_{ds} \approx \dfrac{4 k_B T}{e}$ 的变化，在室温下约为 $100\mathrm{mV}$[171]。

附录F 光学格式和分辨率

表 F.1 光学格式[525]

格式	对角线/mm	H/mm	V/mm	结论
$1/n$ 英寸	$16/n$	$12.8/n$	$9.6/n$	$n<3$
$1/n$ 英寸	$18/n$	$14.4/n$	$10.8/n$	$n\geqslant 3$
35mm	43.27	36.00	24.00	a.r.3:2
APS-C	27.26	22.7	15.1	—
4/3	21.63	17.3	13.0	a.r.4:3

表 F.2 分辨率

缩略词	全称	分辨率
CIF	通用中间格式	352×288
QCIF	1/4 通用中间格式	176×144
VGA	视频图像阵列	640×480
QVGA	1/4 通用中间格式	320×240
SVGA	超级视频图像阵列	800×400
XGA	扩展图像阵列	1024×768
MXGA	超扩展图像阵列	1600×1200

参 考 文 献

[1] A. Moini. *Vision Chips*. Kluwer Acadmic Publisher, Dordrecht, The Netherlands, 2000.

[2] J. Nakamura, editor. *Image Sensors and Signal Processing for Digital Still Cameras*. CRC Press, Boca Raton, FL, 2005.

[3] K. Yonemoto. *Fundamentals and Applications of CCD/CMOS Image Sensors*. CQ Pub. Co., Ltd., Tokyo, Japan, 2003. In Japanese.

[4] O. Yadid-Pecht and R. Etinne-Cummings, editors. *CMOS Imagers: From Phototransduction to Image Processing*. Kluwer Academic Publishers, Dordrecht, The Netherlands, 2004.

[5] Y. Takemura. *CCD Camera Technologies*. Corona Pub., Co. Ltd., Tokyo, Japan, 1997. In Japanese.

[6] Y. Kiuchi. *Fundamentals and Applications of Image Sensors*. The Nikkankogyo Shimbun Ltd., Tokyo, Japan, 1991. In Japanese.

[7] T. Ando and H. Komobuchi. *Introduction of Solid-State Image Sensors*. Nihon Riko Syuppan Kai, Tokyo, Japan, 1999. In Japanese.

[8] A. J. P. Theuwissen. *Solid-State Imaging with Charge-Coupled Devices*. Kluwer Academic Pub., Dordrecht, The Netherlands, 1995.

[9] S. Morrison. A new type of photosensitive junction device. *Solid-State Electron.*, 5: 485-494, 1963.

[10] J. W. Horton, R. V. Mazza, and H. Dym. The scanistor—A solid-state image scanner. *Proc. IEEE*, 52(12): 1513-1528, December 1964.

[11] P. K. Weimer, G. Sadasiv, J. E. Meyer, Jr., L. Meray-Horvath, and W. S. Pike. A self-scanned solid-state image sensor. *Proc. IEEE*, 55(9): 1591-1602, September 1967.

[12] M. A. Schuster and G. Strull. A monolithic mosaic of photon sensors for solidstate imaging applications. *IEEE Trans. Electron Devices*, ED-13(12): 907-912, December 1966.

[13] G. P. Weckler. Operation of p-n Junction Photodetectors in a Photon Flux Integration Mode. *IEEE J. Solid-State Circuits*, SC-2(3): 65-73, September 1967.

[14] R. H. Dyck and G. P. Weckler. Integrated arrays of silicon photodetectors for image sensing. *IEEE Trans. Electron Devices*, ED-15(4): 196-201, April 1968.

[15] P. J. Noble. Self-scanned silicon image detector arrays. *IEEE Trans. Electron Devices*, ED-15(4): 202-209, April 1968.

[16] R. A. Anders, D. E. Callahan, W. F. List, D. H. McCann, and M. A. Schuster. Developmental solid-state imaging system. *IEEE Trans. Electron Devices*, ED-15(4): 191-196, April 1968.

[17] G. Sadasiv, P. K. Weimer, and W. S. Pike. Thin-film circuits for scanning image-sensor arrays. *IEEE Trans. Electron Devices*, ED-15(4): 215-219, April 1968.

[18] W. F. List. Solid-state imaging—Methods of approach. *IEEE Trans. Electron Devices*, ED-15(4): 256-261, April 1968.

[19] W. S. Boyle and G. E. Smith. Charge-Coupled Semiconductor Devices. *Bell System Tech. J.*, 49: 587-593, 1970.

[20] G. F. Amelio, M. F. Tompsett, and G. E. Smith. Experimental Verification of the Charge-Coupled Semiconductor Device Concept. *Bell System Tech. J.*, 49: 593-600, 1970.

[21] G. E. Smith. The invention of the CCD. *Nuclear Instruments & Methods in Physics Research A*, 471: 1-5, 2001.

[22] N. Koike, I. Takemoto, K. Satoh, S. Hanamura, S. Nagahara, and M. Kubo. MOS Area Sensor: Part I—Design Consideration and Performance of an n-p-n Structure 484 * 384 Element Color MOS Imager. *IEEE Trans. Electron Devices*, 27(8): 1682-1687, August 1980.

[23] H. Nabeyama, S. Nagahara, H. Shimizu, M. Noda, and M. Masuda. All Solid State Color Camera with Single-Chip MOS Imager. *IEEE Trans. Consumer Electron.*, CE-27(1): 40-46, February 1981.

[24] S. Ohba, M. Nakai, H. Ando, S. Hanamura, S. Shimada, K. Satoh, K. Takahashi, M. Kubo, and T. Fujita. MOS Area Sensor: Part II—Low-Noise MOS Area Sensor with Antiblooming Photodiodes. *IEEE J. Solid-State Circuits*, 15(4): 747-752, August 1980.

[25] M. Aoki, H. Ando, S. Ohba, I. Takemoto, S. Nagahara, T. Nakano, M. Kubo, and T. Fujita. 2/3-Inch Format MOS Single-Chip Color Imager. *IEEE J. Solid-State Circuits*, 17(2): 375-380, April 1982.

[26] H. Ando, S. Ohba, M. Nakai, T. Ozaki, N. Ozawa, K. Ikeda, T. Masuhara, T. Imaide, I. Takemoto, T. Suzuki, and T. Fujita. Design consideration and performance of a new MOS imaging device. *IEEE Trans. Electron Devices*, ED-32(8): 1484-1489, August 1985.

[27] T. Kinugasa, M. Noda, T. Imaide, I. Aizawa, Y. Todaka, and M Ozawa. An Electronic Variable-Shutter System in Video Camera Use. *IEEE Trans. Consumer Electron.*, CE-33(3): 249-255, August 1987.

[28] K. Senda, S. Terakawa, Y. Hiroshima, and T. Kunii. Analysis of chargepriming transfer efficiency in CPD image sensors. *IEEE Trans. Electron Devices*, 31(9): 1324-1328, September 1984.

[29] T. Nakamura, K. Matsumoto, R. Hyuga, and A Yusa. A new MOS image sensor operating in a nondestructive readout mode. In *Tech. Dig. Int'l Electron Devices Meeting (IEDM)*, pages 353-356, 1986.

[30] J. Hynecek. A new device architecture suitable for high-resolution and highperformance image sensors. *IEEE Trans. Electron Devices*, 35(5): 646-652, May 1988.

[31] N. Tanaka, T. Ohmi, and Y. Nakamura. A novel bipolar imaging device with self-noise-reduction capability. *IEEE Trans. Electron Devices*, 36(1): 31-38, January 1989.

[32] A. Yusa, J. Nishizawa, M. Imai, H. Yamada, J. Nakamura, T. Mizoguchi, Y. Ohta, and M. Takayama. SIT Image sensor: Design considerations and characteristics. *IEEE Trans. Electron Devices*, 33(6): 735-742, June 1986.

[33] F. Andoh, K. Taketoshi, K. Nakamura, and M. Imai. Amplified MOS Intelligent Imager. *J. Inst. Image Information & Television Eng.*, 41(11): 1075-1082, November 1987. In Japanese.

[34] F. Andoh, K. Taketoshi, J. Yamazaki, M. Sugawara, Y. Fujita, K. Mitani, Y. Matuzawa, K. Miyata, and S. Araki. A 250000-pixel image sensor with FET amplification at each pixel for high-speed television cameras. *Dig. Tech. Papers Int'l Solid-State Circuits Conf. (ISSCC)*, pages 212-213, 298, February 1990.

[35] M. Kyomasu. A New MOS Imager Using Photodiode as Current Source. *IEEE J. Solid-State Circuits*, 26(8): 1116-1122, August 1991.

[36] M. Sugawara, H. Kawashima, F. Andoh, N. Murata, Y. Fujita, and M. Yamawaki. An amplified MOS imager suited for image processing. *Dig. Tech. Papers Int'l Solid-State Circuits Conf. (ISSCC)*, pages 228-229, February 1994.

[37] M. Yamawaki, H. Kawashima, N. Murata, F. Andoh, M. Sugawara, and Y. Fujita. A pixel size shrinkage of amplified MOS imager with two-line mixing. *IEEE Trans. Electron Devices*, 1996.

[38] E. R. Fossum. Active pixel sensors—Are CCD's dinosaurs? In *Proc. SPIE*, volume 1900 of *Charge-Coupled Devices and Optical Sensors III*, pages 2-14, 1993.

[39] E. R. Fossum. CMOS Image Sensors: Electronic Camera-On-A-Chip. *IEEE Trans. Electron Devices*, 44(10): 1689-1698, October 1997.

[40] R. M. Guidash, T.-H. Lee, P. P. K. Lee, D. H. Sackett, C. I. Drowley, M. S. Swenson, L. Arbaugh, R.

Hollstein, F. Shapiro, and S. Domer. A 0.6m CMOS pinned photodiode color imager technology. In *Tech. Dig. Int'l Electron Devices Meeting (IEDM)*, pages 927-929, December 1997.

[41] H. S. Wong. Technology and device scaling considerations for CMOS imagers. *IEEE Trans. Electron Devices*, 43(12): 2131-2141, December 1996.

[42] C. Mead. *Analog VLSI and Neural Systems*. Addison-Wesley Publishing Company, Reading, MA, 1989.

[43] C. Koch and H. Liu. *VISION CHIPS, Implementing Vision Algorithms with Analog VLSI Circuits*. IEEE Computer Society, Los Alamitos, CA, 1995.

[44] T. M. Bernard, B. Y. Zavidovique, and F. J. Devos. A Programmbale Artificial Retina. *IEEE J. Solid-State Circuits*, 28(7): 789-798, July 1993.

[45] S. Kameda and T. Yagi. An analog VLSI chip emulating sustained and transient response channels of the vertebrate retina. *IEEE Trans. Neural Networks*, 14(5): 1405-1412, September 2003.

[46] R. Takami, K. Shimonomura, S. Kamedaa, and T. Yagi. An image preprocessing system employing neuromorphic 100 * 100 pixel silicon retina. In *Int'l Symp. Circuits & Systems (ISCAS)*, pages 2771-2774, Kobe, Japan, May 2005.

[47] www.itrs.net/.

[48] G. Moore. Cramming more components onto integrated circuits. *Electronics*, 38(8), April 1965.

[49] L. Fortuna, P. Arena, D. Balya, and A. Zarandy. Cellular neural networks: a paradigm for nonlinear spatio-temporal processing. *IEEE Circuits & Systems Mag.*, 1(4): 6-21, 2001.

[50] S. Espejo, A. Rodríguez-Vázquez, R. Domínguez-Castro, J. L. Huertas, and E. Sánchez-Sinencio. Smart-pixel cellular neural networks in analog currentmode CMOS technology. *IEEE J. Solid-State Circuits*, 29(8): 895-905, August 1994.

[51] R. Domínguez-Castro, S. Espejo, A. Rodríguez-Vázquez, R. A. Carmona, P. Földesy, A. Zárandy, P. Szolgay, T. Szirányi, and T. Roska. A 0.8-μm CMOS two-dimensional programmablemixed-signal focal-plane array processor with on-chip binary imaging and instructions storage. *IEEE J. Solid-State Circuits*, 32(7): 1013-1026, July 1997.

[52] G. Liñan-Cembrano, L. Canranza, S. Espejo, R. Dominguez-Castro, and A. Rodíguez-Vázquez. CMOS mixed-signal flexible vision chips. In M. Valle, editor, *Smart Adaptive Systems on Silicon*, pages 103-118. Kluwer Academic Pub., Dordecht, The Netherlands, 2004.

[53] R. Etienne-Cummings, Z. K. Kalayjian, and D. Cai. A Programmable Focal-Plane MIMD Image Processor Chip. *IEEE J. Solid-State Circuits*, 36(1): 64-73, January 2001.

[54] E. Culurciello, R. Etienne-Cummings, and K. A. Boahen. A Biomorphic Digital Image Sensor. *IEEE J. Solid-State Circuits*, 38(2): 281-294, February 2003.

[55] P. Dudek and P. J. Hicks. A General-Purpose Processor-per-Pixel Analog SIMD Vision Chip. *IEEE Trans. Circuits & Systems I*, 52(1): 13-20, January 2005.

[56] P. Dudek and P. J. Hicks. A CMOS General-Purpose Sampled-Data Analog Processing Element. *IEEE Trans. Circuits & Systems II*, 47(5): 467-473, May 2000.

[57] M. Ishikawa, K. Ogawa, T. Komuro, and I. Ishii. A CMOS vision chip with SIMD processing element array for 1 ms image processing. In *Dig. Tech. Papers Int'l Solid-State Circuits Conf. (ISSCC)*, pages 206-207, February 1999.

[58] M. Ishikawa and T. Komuro. Digital vision chips and high-speed vision systems. In *Dig. Tech. Papers Symp. VLSI Circuits*, pages 1-4, June 2001.

[59] N. Mukohzaka, H. Toyoda, S. Mizuno, M. H. Wu, Y. Nakabo, and M. Ishikawa. Column parallel vision system: CPV. In *Proc. SPIE*, volume 4669, pages 21-28, San Jose, CA, January 2002.

[60] T. Komuro, S. Kagami, and M. Ishikawa. A high speed digital vision chip with multi-grained parallel

processing capability. In *IEEE Workshop on Charge-Coupled Devices & Advanced Image Sensors*, Elmau, Geramany, June 2003.

[61] T. Komuro, I. Ishii, M. Ishikawa, and A. Yoshida. A Digital Vision Chip Specialized for High-Speed Target Tracking. *IEEE Trans. Electron Devices*, 50(1): 191-199, January 2003.

[62] T. Komuro, S. Kagami, and M. Ishikawa. A dynamically reconfigurable SIMD processor for a vision chip. *IEEE J. Solid-State Circuits*, 39(1): 265-268, January 2004.

[63] T. Komuro, S. Kagami, M. Ishikawa, and Y. Katayama. Development of a Bit-level Compiler for Massively Parallel Vision Chips. In *IEEE Int'l Workshop Computer Architecture for Machine Perception (CAMP)*, pages 204-209, Palermo, July 2005.

[64] D. Renshaw, P. B. Denyer, G. Wang, and M. Lu. ASIC VISION. In *Proc. Custom Integrated Circuits Conf. (CICC)*, pages 7.3/1-7.3/4, May 1990.

[65] P. B. Denyer, D. Renshaw, Wang G., and Lu M. CMOS image sensors for multimedia applications. In *Proc. Custom Integrated Circuits Conf. (CICC)*, pages 11.5.1-11.5.4, May 1993.

[66] K. Chen, M. Afghani, P. E. Danielsson, and C. Svensson. PASIC: A processor-A/D converter-sensor integrated circuit. In *Int'l Symp. Circuits & Systems (ISCAS)*, volume 3, pages 1705-1708, May 1990.

[67] J.-E. Eklund, C. Svensson, and A. Åström. VLSI Implementation of a Focal Plane Image Processor—A Realization of the Near-Sensor Image Processing Concept. *IEEE Trans. VLSI Systems*, 4(3): 322-335, September 1996.

[68] L. Lindgren, J. Melander, R. Johansson, and B. Moller. A multiresolution 100-GOPS 4-Gpixels/s programmable smart vision sensor for multisense imaging. *IEEE J. Solid-State Circuits*, 40(6): 1350-1359, June 2005.

[69] E. D. Palik. *Handbook of Optical Constants of Solids*. Academic Press, New York, NY, 1977.

[70] S. E. Swirhun, H.-H. Kwark, and R. M. Swanson. Measurement of electron lifetime, electron mobility and band-gap narrowing in heavily doped p-type silicon. In *Tech. Dig. Int'l Electron Devices Meeting (IEDM)*, pages 24-27, 1986.

[71] J. del Alamo, S. Swirhun, and R. M. Swanson. Measuring and modeling minority carrier transport in heavily doped silicon. *Solid-State Electronic.*, 28: 47-54, 1985.

[72] J. del Alamo, S. Swirhun, and R. M. Swanson. Simultaneous measurement of hole lifetime, hole mobility and bandgap narrowing in heavily doped n-type silicon. In *Tech. Dig. Int'l Electron Devices Meeting (IEDM)*, pages 290-293, 1985.

[73] J. S. Lee, R. I. Hornsey, and D. Renshaw. Analysis of CMOS Photodiodes. I. Quantum efficiency. *IEEE Trans. Electron Devices*, 50(5): 1233-1238, May 2003.

[74] J. S. Lee, R. I. Hornsey, and D. Renshaw. Analysis of CMOS Photodiodes. II. Lateral photoresponse. *IEEE Trans. Electron Devices*, 50(5): 1239-1245, May 2003.

[75] S. G. Chamberlain, D. J. Roulston, and S. P. Desai. Spectral Response LimitationMechanisms of a Shallow Junction n^+-p Photodiode. *IEEE J. Solid-State Circuits*, SC-13(1): 167-172, February 1978.

[76] I. Murakami, T. Nakano, K. Hatano, Y. Nakashiba, M. Furumiya, T. Nagata, H. Utsumi, S. Uchida, K. Arai, N. Mutoh, A. Kohno, N. Teranishi, and Y. Hokari. New Technologies of Photo-Sensitivity Improvement and VOD Shutter Voltage Reduction for CCD Image Sensors. In *Proc. SPIE*, volume 3649, pages 14-21, San Jose, CA, 1999.

[77] S. M. Sze. *Physics of Semiconductor Devices*. John Wiley & Sons, Inc., New York, NY, 1981.

[78] N. V. Loukianova, H. O. Folkerts, J. P. V. Maas, D. W. E. Verbugt, A. J. Mierop, W. Hoekstra, E. Roks, and A. J. P. Theuwissen. Leakage current modeling of test structures for characterization of

dark current in CMOS image sensors. *IEEE Trans. Electron Devices*, 50(1): 77 83, January 2003.

[79] B. Pain, T. Cunningham, B. Hancock, C. Wrigley, and C. Sun. Excess Noise and Dark Current Mechanism in CMOS Imagers. In *IEEE Workshop on Charge-Coupled Devices & Advanced Image Sensors*, pages 145-148, Karuizawa, Japan, June 2005.

[80] A. S. Grove. *Physics and Technology of Semiconductor Devices*. John Wiley & Sons, Inc., New York, NY, 1967.

[81] G. A. M. Hurkx, H. C. de Graaff, W. J. Kloosterman, and M. P. G. Knuvers. A new analytical diode model including tunneling and avalanche breakdown. *IEEE Trans. Electron Devices*, 39(9): 2090-2098, September 1992.

[82] H. O. Folkerts, J. P. V. Maas, D. W. E. Vergugt, A. J. Mierop, W. Hoekstra, N. V. Loukianova, and E. Rocks. Characterization of Dark Current in CMOS Image Sensors. In *IEEE Workshop on Charge-Coupled Devices & Advanced Image Sensors*, Elmau, Germany, May 2003.

[83] H. Zimmermann. *Silicon Optoelectronic Integrated Cicuits*. Springer-Verlag, Berlin, Germany, 2004.

[84] S. Radovanović, A.-J. Annema, and B. Nauta. *High-Speed Photodiodes in Standard CMOS Technology*. Springer, Dordecht, The Netherlands, 2006.

[85] B. Razavi. *Design of Integrated Circuits for Optical Communications*. McGraw-Hill Companies Inc., New York, NY, 2003.

[86] J. Singh. *SemiconductorOptoelectronics: Physics and Technology*. McGraw-Hill, Inc., New York, NY, 1995.

[87] V. Brajovic, K. Mori, and N. Jankovic. 100frames/s CMOS Range Image Sensor. In *Dig. Tech. Papers Int'l Solid-State Circuits Conf. (ISSCC)*, pages 256-257, February 2001.

[88] Y. Takiguchi, H. Maruyama, M. Kosugi, F. Andoh, T. Kato, K. Tanioka, J. Yamazaki, K. Tsuji, and T. Kawamura. A CMOS Imager Hybridized to an Avalanche Multiplied Film. *IEEE Trans. Electron Devices*, 44(10): 1783-1788, October 1997.

[89] A. Biber, P. Seitz, and H. Jäckel. Avalanche Photodiode Image Sensor in Standard BiCMOS Technology. *IEEE Trans. Electron Devices*, 47(11): 2241-2243, November 2000.

[90] J. C. Jackson, P. K. Hurley, A. P. Morrison, B. Lane, and A. Mathewson. Comparing Leakage Currents and Dark Count Rates in Shallow Junction Geiger-Mode Avalanche Photodiodes. *Appl. Phys. Lett.*, 80(22): 4100-4102, June 2002.

[91] J. C. Jackson, A. P. Morrison, D. Phelan, and A. Mathewson. A Novel Silicon Geiger-Mode Avalanche Photodiode. In *Tech. Dig. Int'l Electron Devices Meeting (IEDM)*, December 2002.

[92] J. C. Jackson, D. Phelan, A. P. Morrison, M. Redfern, and A. Mathewson. Towards integrated single photon counting arrays. *Opt. Eng.*, 42(1): 112-118, January 2003.

[93] A. M. Moloney, A. P. Morrison, J. C. Jackson, A. Mathewson, J. Alderman, J. Donnelly, B. O'Neill, A.-.M. Kelleher, G. Healy, and P. J. Murphy. Monolithically Integrated Avalanche Photodiode and Transimpedance Amplifier in a Hybrid Bulk/SOI CMOS Process. *Electron. Lett.*, 39(4): 391-392, February 2003.

[94] J. C. Jackson, J. Donnelly, B. O'Neill, A-M. Kelleher, G. Healy, A. P. Morrison, and A. Mathewson. Integrated Bulk/SOI APD Sensor: Bulk Substrate Inspection with Geiger-Mode Avalanche Photodiodes. *Electron. Lett.*, 39(9): 735-736, May 2003.

[95] A. Rochas, A. R. Pauchard, P.-A. Besse, D. Pantic, Z. Prijic, and R. S. Popovic. Low-Noise Silicon Avalanche Photodiodes Fabricated in Conventional CMOS Technologies. *IEEE Trans. Electron Devices*, 49(3): 387-394, March 2002.

[96] A. Rochas, M. Gosch, A. Serov, P. A. Besse, R. S. Popovic, T. Lasser, and R. Rigler. First Fully Integrated 2-D Array of Single-Photon Detectors in Standard CMOS Technology. *IEEE Photon.*

Tech. Lett. ,15(7): 963-965,July 2003.

[97] C. Niclass,A. Rochas, P. -A. Besse, and E. Charbon. Toward a 3-D Camera Based on Single Photon Avalanche Diodes. *IEEE Selcted Topic Quantum Electron.* ,2004.

[98] C. Niclass,A. Rochas, P. -A. Besse, and E. Charbon. Design and Characterization of a CMOS 3-D Image Sensor Based on Single Photon Avalanche Diodes. *IEEE J. Solid-State Circuits*,40(9): 1847-1854,September 2005.

[99] S. Bellis,R. Wilcock, and C. Jackson. Photon Counting Imaging: the DigitalAPD. In *SPIE-IS&T Electronic Imaging: Sensors,Cameras, and Systems for Scientific/Industrial Applications VII*, volume 6068,pages 60680D-1-D-10,San Jose,CA,January 2006.

[100] C. J. Stapels,W. G. Lawrence, F. L. Augustine, and J. F. Christian. Characterization of a CMOS Geiger Photodiode Pixel. *IEEE Trans. Electron Devices*,53(4): 631-635,April 2006.

[101] H. Finkelstein,M. J. Hsu,and S. C. Esener. STI-Bounded Single-Photon Avalanche Diode in a Deep-Submicrometer CMOS Technology. *IEEE Electron Device Lett.* ,27(11): 887-889,November 2006.

[102] M. Kubota,T. Kato,S. Suzuki, H. Maruyama,K. Shidara, K. Tanioka, K. Sameshima, T. Makishima, K. Tsuji,, T. Hirai, and T. Yoshida. Ultra highsensitivity newsuper-HARP camera. *IEEE Trans. Broadcast.* ,42(3): 251-258,September 1996.

[103] T. Watabe,M. Goto, H. Ohtake, H. Maruyama, M. Abe, K. Tanioka, and N. Egami. New signal readout method for ultra high-sensitivity CMOS image sensor. *IEEE Trans. Electron Devices*, 50(1): 63-69,January 2003.

[104] S. Aihara, Y. Hirano, T. Tajima, K. Tanioka, M. Abe, N. Saito, N. Kamata, and D. Terunuma. Wavelength selectivities of organic photoconductive films: Dye-doped polysilanes and zinc phthalocyanine/tris-8-hydroxyquinoline aluminum double layer. *Appl. Phys. Lett.* ,82(4): 511-513, January 2003.

[105] T. Watanabe,S. Aihara, N. Egami, M. Kubota, K. Tanioka, N. Kamata, and D. Terunuma. CMOS Image Sensor Overlaid with an Organic Photoconductive Film. In *IEEE Workshop on Charge-Coupled Devices & Advanced Image Sensors*,pages 48-51,Karuizawa,Japan,June 2005.

[106] S. Takada, M. Ihama, and M. Inuiya. CMOS Image Sensor with Organic Photoconductive Layer Having Narrow Absorption Band and Proposal of Stack Type Solid-State Image Sensors. In *Proc. SPIE*,volume 6068,pages 60680A-1-A8,San Jose,CA,January 2006.

[107] J. Burm,K. I. Litvin,D. W. Woodard,W. J. Schaff, P. Mandeville, M. A. Jaspan, M. M. Gitin, and L. F. Eastman. High-frequency, high-efficiency MSM photodetectors. *IEEE J. Quantum Electron.* , 31(8): 1504-1509,August 1995.

[108] K. Tanaka,F. Ando,K. Taketoshi, I. Ohishi, and G. Asari. Novel Digital Photosensor Cell in GaAs IC Using Conversion of Light Intensity to Pulse Frequency. *Jpn. J. Appl. Phys.* ,32(11A): 5002-5007,November 1993.

[109] E. Lange, E. Funatsu, J. Ohta, and K. Kyuma. Direct image processing using arrays of variable-sensitivity photodetectors. In *Dig. Tech. Papers Int'l Solid-State Circuits Conf. (ISSCC)*, pages 228-229,February 1995.

[110] H. -B. Lee, H. -S. An, H. -I. Cho, J. -H. Lee, and S. -H. Hahm. UV photoresponsive characteristics of an n-channel GaN Schottky-barrier MISFET forUV image sensors. *IEEE Electron Device Lett.* , 27(8): 656-658,August 2006.

[111] M. Abe. Image sensors and circuit technologies. In T. Enomoto, editor,*Video/Image LSI System Design Technology*,pages 208-248. Corona Pub. ,Co. Ltd. ,Tokyo,Japan,2003. In Japanese.

[112] N. Teranishi,A. Kohono,Y. Ishihara, E. Oda, and K. Arai. No image lag photodiode structure in the interline CCD image sensor. In *Tech. Dig. Int'l Electron Devices Meeting (IEDM)*, pages 324-

327, 1982.

[113] B. C. Burkey, W. C. Chang, J. Littlehale, T. H. Lee, T. J. Tredwell, J. P. Lavine, and E. A. Trabka. The pinned photodiode for an interline-transfer CCD image sensor. In *Tech. Dig. Int'l Electron Devices Meeting (IEDM)*, pages 28-31, 1984.

[114] I. Inoue, H. Ihara, H. Yamashita, T. Yamaguchi, H. Nozaki, and R. Miyagawa. Low dark current pinned photo-diode for CMOS image sensor. In *IEEE Workshop on Charge-Coupled Devices & Advanced Image Sensors*, pages 25-32, Karuizawa, Japan, June 1999.

[115] S. Agwani, R Cichomski, M. Gorder, A. Niederkorn, Sknow M., and K. Wanda. A 1/3" VGA CMOS Imaging System On a Chip. In *IEEE Workshop on Charge-Coupled Devices & Advanced Image Sensors*, pages 21-24, Karuizawa, Japan, June 1999.

[116] K. Yonemoto and H. Sumi. A CMOS image sensor with a simple fixedpattern-noise-reduction technology and a hole accumulation diode. *IEEE J. Solid-State Circuits*, 2000.

[117] M. Noda, T. Imaide, T. Kinugasa, and R. Nishimura. A Solid State Color Video Camera with a Horizontal Readout MOS Imager. *IEEE Trans. Consumer Electron.*, CE-32(3): 329-336, August 1986.

[118] S. Miyatake, M. Miyamoto, K. Ishida, T. Morimoto, Y. Masaki, and H. Tanabe. Transversal-readout architecture for CMOS active pixel image sensors. *IEEE Trans. Electron Devices*, 50(1): 121-129, January 2003.

[119] J. D. Plummer and J. D. Meindl. MOS electronics for a portable reading aid for the blind. *IEEE J. Solid-State Circuits*, 7(2): 111-119, April 1972.

[120] L. J. Kozlowski, J. Luo, W. E. Kleinhans, and T. Liu. Comparison of Passive and Active Pixel Schemes for CMOS Visible Imagers. In *Proc. SPIE*, volume 3360 of *Infrared Readout Electronics IV*, pages 101-110, Orland, FL, April 1998.

[121] Y. Endo, Y. Nitta, H. Kubo, T. Murao, K. Shimomura, M. Kimura, K. Watanabe, and S. Komori. 4-micron pixel CMOS iamge sensor with low image lag and high-temperature operability. In *Proc. SPIE*, volume 5017, pages 196-204, Santa Clara, CA, January 2003.

[122] I. Inoue, N. Tanaka, H. Yamashita, T. Yamaguchi, H. Ishiwata, and H. Ihara. Low-leakage-current and low-operating-voltage buried photodiode for a CMOS imager. *IEEE Trans. Electron Devices*, 50(1): 43-47, January 2003.

[123] O. Yadid-Pecht, B. Pain, C. Staller, C. Clark, and E. Fossum. CMOS active pixel sensor star tracker with regional electronic shutter. *IEEE J. Solid-State Circuits*, 32(2): 285-288, February 1997.

[124] S. E. Kemeny, R. Panicacci, B. Pain, L. Matthies, and E. R. Fossum. Multiresolution Image Sensor. *IEEE Trans. Circuits & Systems Video Tech.*, 7(4): 575-583, August 1997.

[125] K. Salama and A. El Gamal. Analysis of active pixel sensor readout circuit. *IEEE Trans. Circuits & Systems I*, 50(7): 941-945, July 2003.

[126] B. Razavi. *Design of Analog CMOS Integrated Circuits*. McGraw-Hill Companies, Inc., New York, NY, 2001.

[127] J. Hynecek. Analysis of the photosite reset in FGA image sensors. *IEEE Trans. Electron Devices*, 37(10): 2193-2200, October 1990.

[128] S. Mendis, S. E. Kemeny, and E. R. Fossum. CMOS active pixel image sensor. *IEEE Trans. Electron Devices*, 41(3): 452-453, March 1994.

[129] R. H. Nixon, S. E. Kemeny, B. Pain, C. O. Staller, and E. R. Fossum. 256 * 256 CMOS active pixel sensor camera-on-a-chip. *IEEE J. Solid-State Circuits*, 31(12): 2046-2050, December 1996.

[130] T. Sugiki, S. Ohsawa, H. Miura, M. Sasaki, N. Nakamura, I. Inoue, M. Hoshino, Y. Tomizawa, and T. Arakawa. A 60mW 10 b CMOS image sensor with column-to-column FPN reduction. In *Dig.*

Tech. Papers Int'l Solid-State Circuits Conf. (ISSCC), pages 108-109, February 2000.

[131] M. F. Snoeij, A. J. P. Theuwissen, K. A. A. Makinwa, and J. H. Huijsing. A CMOS Imager with Column-Level ADC Using Dynamic Column Fixed-Pattern Noise Reduction. *IEEE J. Solid-State Circuits*, 41(12): 3007-3015, December 2006.

[132] M. Furuta, S. Kawahito, T. Inoue, and Y. Nishikawa. A cyclic A/D converter with pixel noise and column-wise offset canceling for CMOS image sensors. In *Proc. European Solid-State Circuits Conf. (ESSCIRC)*, pages 411-414, Grenoble, France, September 2005.

[133] M. Waeny, S. Tanner, S. Lauxtermann, N. Blanc, M. Willemin, M. Rechsteiner, E. Doering, J. Grupp, P. Seitz, F. Pellandini, and M. Ansorge. High sensitivity and high dynamic, digital CMOS imager. In *Proc. SPIE*, volume 406, pages 78-84, May 2001.

[134] S. Smith, J. Hurwitz, M. Torrie, D. Baxter, A. Holmes, M. Panaghiston, R. Henderson, A. Murray, S. Anderson, and P. Denyer. A single-chip 306 * 244-pixel CMOS NTSC video camera. In *Dig. Tech. Papers Int'l Solid-State Circuits Conf. (ISSCC)*, pages 170-171, February 1998.

[135] M. J. Loinaz, K. J. Singh, A. J. Blanksby, D. A. Inglis, K. Azadet, and B. D. Ackland. A 200-mW, 3. 3-V, CMOS color camera IC producing 352 * 288 24-b video at 30 frames/s. *IEEE J. Solid-State Circuits*, 33(12): 2092-2103, December 1998.

[136] Z. Zhou, B. Pain, and E. E. Fossum. CMOS active pixel sensor with on-chip successive approximation analog-to-digital converter. *IEEE Trans. Electron Devices*, 44(10): 1759-1763, October 1997.

[137] I. Takayanagi, M. Shirakawa, K. Mitani, M. Sugawara, S. Iversen, J. Moholt, J. Nakamura, and E. R. Fossum. 1 1/4 inch 8. 3M pixel digital output CMOS APS for UDTV application. In Dig. Tech. Papers Int'l Solid-State Circuits Conf. (ISSCC), pages 216-217, February 2003.

[138] K. Findlater, R. Henderson, D. Baxter, J. E. D. Hurwitz, L. Grant, Y. Cazaux, F. Roy, D. Herault, and Y. Marcellier. SXGA pinned photodiodeCMOS image sensor in 0. 35μm technology. In *Dig. Tech. Papers Int'l Solid-State Circuits Conf. (ISSCC)*, page 218, February 2003.

[139] S. Decker, D. McGrath, K. Brehmer, and C. G. Sodini. A 256 * 256 CMOS imaging array with wide dynamic range pixels and column-parallel digital output. *IEEE J. Solid-State Circuits*, 33(12): 2081-2091, December 1998.

[140] D. Yang, B. Fowler, and A. El Gamal. A Nyquist Rate Pixel Level ADC for CMOS Image Sensors. *IEEE J. Solid-State Circuits*, 34(3): 348-356, March 1999.

[141] F. Andoh, H. Shimamoto, and Y. Fujita. A Digital Pixel Image Sensor for Real-Time Readout. *IEEE Trans. Electron Devices*, 47(11): 2123-2127, November 2000.

[142] M. Willemin, N. Blanc, G. K. Lang, S. Lauxtermann, P. Schwider, P. Seitz, and M. Wäny. Optical characterizationmethods for solid-state image sensors. *Optics and Lasers Eng.*, 36(2): 185-194, 2001.

[143] G. R. Hopkinson, T. M. Goodman, and S. R. Prince. *A guide to the use and calibration of detector arrya equipment*. SPIE Press, Bellingham, Washington, 2004.

[144] M. F. Snoeij, A. Theuwissen, K. Makinwa, and J. H. Huijsing. A CMOS Imager with Column-Level ADC Using Dynamic Column FPN Reduction. In *Dig. Tech. Papers Int'l Solid-State Circuits Conf. (ISSCC)*, pages 2014-2023, February 2006.

[145] J. E. Carnes and W. F. Kosonocky. Noise source in charge-coupled devices. *RCA Review*, 33(2): 327-343, June 1972.

[146] B. Pain, G. Yang, M. Ortiz, C. Wrigley, B. Hancock, and T. Cunningham. Analysis and enhancement of low-light-level performance of photodiode-type CMOS active pixel imagers operated with sub-threshold reset. In *IEEE Workshop on Charge-Coupled Devices & Advanced Image Sensors*, pages 140-143, Karuizawa, Japan, June 1999.

[147] B. Pain, G. Yang, T. J. Cunningham, C. Wrigley, and B. Hancock. An Enhanced-Performance CMOS Imager with a Flushed-Reset Photodiode Pixel. *IEEE Trans. Electron Devices*, 50(1): 48-56, January 2003.

[148] B. E. Bayer. Color imaging array. US patent 3,971,065, July 1976.

[149] M. Kasano, Y. Inaba, M. Mori, S. Kasuga, T. Murata, and T. Yamaguchi. A 2.0-μm Pixel Pitch MOS Image Sensor with 1.5 Transistor/Pixel and an Amorphous Si Color Filter. *IEEE Trans. Electron Devices*, 53(4): 611-617, April 2006.

[150] R. D. McGrath, H. Fujita, R. M. Guidash, T. J. Kenney, and W. Xu. Shared pixels for CMOS image sensor arrays. In *IEEE Workshop on Charge-Coupled Devices & Advanced Image Sensors*, pages 9-12, Karuizawa, Japan, June 2005.

[151] H. Takahashi, M. Kinoshita, K. Morita, T. Shirai, T. Sato, T. Kimura, H. Yuzurihara, and S. Inoue. A 3.9μm pixel pitch VGA format 10 b digital image sensor with 1.5-transistor/pixel. In *Dig. Tech. Papers Int'l Solid-State Circuits Conf. (ISSCC)*, pages 108-109, February 2004.

[152] M. Mori, M. Katsuno, S. Kasuga, T. Murata, and T. Yamaguchi. A 1/4in 2M pixel CMOS image sensor with 1.75 transistor/pixel. In *Dig. Tech. Papers Int'l Solid-State Circuits Conf. (ISSCC)*, pages 110-111, February 2004.

[153] K. Mabuchi, N. Nakamura, E. Funatsu, T. Abe, T. Umeda, T. Hoshino, R. Suzuki, and H. Sumi. CMOS image sensor using a floating diffusion driving buried photodiode. In *Dig. Tech. Papers Int'l Solid-State Circuits Conf. (ISSCC)*, pages 112-113, February 2004.

[154] M. Murakami, M. Masuyama, S. Tanaka, M. Uchida, K. Fujiwara, M. Kojima, Y. Matsunaga, and S. Mayumi. 2.8μm-Pixel Image Sensor Maicovicon™. In *IEEE Workshop on Charge-Coupled Devices & Advanced Image Sensors*, pages 13-14, Karuizawa, Japan, June 2005.

[155] Y. C. Kim, Y. T. Kim, S. H. Choi, H. K. Kong, S. I. Hwang, J. H. Ko, B. S. Kim, T. Asaba, S. H. Lim, J. S. Hahn, J. H. Im, T. S. Oh, D. M. Yi, J. M. Lee, W. P. Yang, J. C. Ahn, E. S. Jung, and Y. H. Lee. 1/2-inch 7.2M Pixel CMOS Image Sensor with 2.25μm Pixels Using 4-Shared Pixel Structure for Pixel-Level Summation. In *Dig. Tech. Papers Int'l Solid-State Circuits Conf. (ISSCC)*, pages 1994-2003, February 2006.

[156] S. Yoshihara, M. Kikuchi, Y. Ito, Y. Inada, S. Kuramochi, H. Wakabayashi, M. Okano, K. Koseki, H. Kuriyama, J. Inutsuka, A. Tajima, T. Nakajima, Y. Kudoh, F. Koga, Y. Kasagi, S. Watanabe, and T. Nomoto T. A 1/1.8-inch 6.4M Pixel 60 frames/s CMOS Image Sensor with Seamless Mode Change. In *Dig. Tech. Papers Int'l Solid-State Circuits Conf. (ISSCC)*, pages 1984-1993, February 2006.

[157] S. Yoshimura, T. Sugiyama, K. Yonemoto, and K. Ueda. A 48 kframe/s CMOS image sensor for real-time 3-D sensing and motion detection. In *Dig. Tech. Papers Int'l Solid-State Circuits Conf. (ISSCC)*, pages 94-95, February 2001.

[158] J. Nakamura, B. Pain, T. Nomoto, T. Nakamura, and Eric R. Fossum. On-Focal-Plane Signal Processing for Current-Mode Active Pixel Sensors. *IEEE Trans. Electron Devices*, 44(10): 1747-1758, October 1997.

[159] Y. Huanga and R. I. Hornsey. Current-mode CMOS image sensor using lateral bipolar phototransistors. *IEEE Trans. Electron Devices*, 50(12): 2570-2573, December 2003.

[160] M. A. Szelezniak, G. W. Deptuch, F. Guilloux, S. Heini, and A. Himmi. Current Mode Monolithic Active Pixel Sensor with Correlated Double Sampling for Charged Particle Detection. *IEEE Sensors Journal*, 7(1): 137-150, January 2007.

[161] L. G. McIlrath, V. S. Clark, P. K. Duane, R. D. McGrath, and W. D. Waskurak. Design and Analysis of a 512 * 768 Current-Mediated Active Pixel Array Image Sensor. *IEEE Trans. Electron Devices*,

44(10),1997.

[162] F. Boussaid, A. Bermak, and A. Bouzerdoum. An ultra-low power operating technique for Megapixels current-mediated CMOS imagers. *IEEE Trans. Consumer Electron.*, 50(1): 46-53, February 2004.

[163] D. Scheffer, B. Dierickx, and G. Meynants. Random Addressable 2048 * 2048 Active Pixel Image Sensor. *IEEE Trans. Electron Devices*, 44(10): 1716-1720, October 1997.

[164] S. Kavadias, B. Dierickx, D. Scheffer, A. Alaerts, D. Uwaerts, and J. Bogaerts. A logarithmic response CMOS image sensor with on-chip calibration. *IEEE J. Solid-State Circuits*, 35(8): 1146-1152, August 2000.

[165] M. Loose, K. Meier, and J. Schemmel. A self-calibrating single-chip CMOS camera with logarithmic response. *IEEE J. Solid-State Circuits*, 36(4): 586-596, April 2001.

[166] Y. Oike, M. Ikeda, and K. Asada. High-Sensitivity and Wide-Dynamic-Range Position Sensor Using Logarithmic-Response and Correlation Circuit. *IEICE Trans. Electron.*, E85-C(8): 1651-1658, August 2002.

[167] L.-W. Lai, C.-H. Lai, and Y.-C. King. A Novel Logarithmic Response CMOS Image Sensor with High Output Voltage Swing and In-Pixel Fixed-Pattern Noise Reduction. *IEEE Sensors Journal*, 4(1): 122-126, February 2004.

[168] B. Choubey, S. Aoyoma, S. Otim, D. Joseph, and S. Collins. An Electronic-Calibration Scheme for Logarithmic CMOS Pixels. *IEEE Sensors Journal*, 6(4): 950-956, August 2006.

[169] T. Kakumoto, S. Yano, M. Kusudaa, K. Kamon, and Y. Tanaka. Logarithmic conversion CMOS image sensor with FPN cancellation and integration circuits. *J. Inst. Image Information & Television Eng.*, 57(8): 1013-1018, August 2003.

[170] J. Lazzaro, S. Ryckebuscha, M. A. Mahowald, and C. A. Mead. Winner-Take-All Networks of $O(n)$ Complexity. In D. Tourestzky, editor, *Advances in Neural Information Processing Systems*, volume 1, pages 703-711. Morgan Kaufmann, San Mateo, CA, 1988.

[171] S.-C. Liu, J. Karmer, G. Indiveri, T. Delbrük, and R. Douglas. *Analog VLSI Cicuits and Principles*. The MIT Press, Cambridge, MA, 2002.

[172] B. K. P. Horn. *Robot Vision*. The MIT Press, Cambridge, MA, 1986.

[173] E. Funatsu, Y. Nitta, Y. Miyake, T. Toyoda, J. Ohta, and K. Kyuma. An Artificial Retina Chip with Current-Mode Focal Plane Image Processing Functions. *IEEE Trans. Electron Devices*, 44(10): 1777-1782, October 1997.

[174] E. Funatsu, Y. Nitta, J. Tanaka, and K. Kyuma. A 128 * 128 Pixel Artificial Retina LSI with Two-Dimensional Filtering Functions. *Jpn. J. Appl. Phys.*, 38(8B): L938-L940, August 1999.

[175] D. Marr. *Vision: A Computational Investigation into the Human Representation and Processing of Visual Information*. W. H. Freeman, 1983.

[176] X. Liu and A. El Gamal. Photocurrent Estimation for a Self-Reset CMOS Image Sensor. In *Proc. SPIE*, volume 4669, pages 304-312, San Jose, CA, 2002.

[177] K. P. Frohmader. A novel MOS compatible light intensity-to-frequency converter suited for monolithic integration. *IEEE J. Solid-State Circuits*, 17(3): 588-591, June 1982.

[178] R. Müller. I2/L timing circuit for the 1 ms-10 s range. *IEEE J. Solid-State Circuits*, 12(2): 139-143, April 1977.

[179] V. Brajovic and T. Kanade. A sorting image sensor: An example of massively parallel intensity-to-time processing for low-latency computational sensors. In *Proc. IEEE Int'l Conf. Robotics & Automation*, pages 1638-1643, Minneapolis, MN, April 1996.

[180] M. Nagata, J. Funakoshi, and A. Iwata. A PWM Signal Processing Core Circuit Based on a Switched

Current Integration Technique. *IEEE J. Solid-State Circuits*, 1998.

[181] M. Nagata, M. Homma, N. Takeda, T. Morie, and A. Iwata. A smart CMOS imager with pixel level PWM signal processing. In *Dig. Tech. Papers Symp. VLSI Circuits*, pages 141-144, June 1999.

[182] M. Shouho, K. Hashiguchi, K. Kagawa, and J. Ohta. A Low-Voltage Pulse-Width-Modulation Image Sensor. In *IEEE Workshop on Charge-Coupled Devices & Advanced Image Sensors*, pages 226-229, Karuizawa, Japan, June 2005.

[183] S. Shishido, I. Nagahata, T. Sasaki, K. Kagawa, M. Nunoshita, and J. Ohta. Demonstration of a low-voltage three-transistor-per-pixel CMOS imagerbased on a pulse-width-modulation readout scheme employed with a onetransistor in-pixel comparator. In *Proc. SPIE*, San Jose, CA, 2007. Electronic Imaging.

[184] D. Yang, A. El Gamal, B. Fowler, and H. Tian. A 640 * 512 CMOS Image Sensor with Ultrawide Dynamic Range Floating-Point Pixel-Level ADC. *IEEE J. Solid-State Circuits*, 34(12): 1821-1834, December 1999.

[185] S. Kleinfelder, S.-H. Lim, X. Liu, and A. El Gamal. A 10 000 Frames/s CMOS Digital Pixel Sensor. *IEEE J. Solid-State Circuits*, 36(12): 2049-2059, December 2001.

[186] W. Biderman, A. El Gamal, S. Ewedemi, J. Reyneri, H. Tian, D. Wile, and D. Yang. A 0.18m High Dynamic Range NTSC/PAL Imaging Systemon-Chip with Embedded DRAM Frame Buffer. In *Dig. Tech. Papers Int'l Solid-State Circuits Conf. (ISSCC)*, pages 212-213, 2003.

[187] J. G. Nicholls, A. R. Martin, B. G. Wallace, and P. A. Fuchs. *From Neuro to Brain*. Sinauer Associates, Inc., Sunderland, MA, 4th edition, 2001.

[188] W. Maass and C. M. Bishop, editors. *Pulsed Neural Networks*. The MIT Press, Cambridge, MA, 1999.

[189] T. Lehmann and R. Woodburn. Biologically-inspired learning in pulsed neural networks. In G. Cauwenberghs and M. A. Bayoumi, editors, *Learning on silicon: adaptive VLSI neural systems*, pages 105-130. Kluwer Academic Pub., Norwell, MA, 1999.

[190] K. Kagawa, K. Yasuoka, D. C. Ng, T. Furumiya, T. Tokuda, J. Ohta, and M. Nunoshita. Pulse-domain digital image processing for vision chips employing low-voltage operation in deep-submicron technologies. *IEEE Selcted Topic Quantum Electron.*, 10(4): 816-828, July 2004.

[191] T. Hammadou. Pixel Level Stochastic Arithmetic for Intelligent Image Capture. In *Proc. SPIE*, volume 5301, pages 161-167, San Jose, CA, January 2004.

[192] W. Yang. A wide-dynamic-range, low-power photosensor array. In *Dig. Tech. Papers Int'l Solid-State Circuits Conf. (ISSCC)*, pages 230-231, February 1994.

[193] K. Kagawa, S. Yamamoto, T. Furumiya, T. Tokuda, M. Nunoshita, and J. Ohta. A pulse-frequency-modulation vision chip using a capacitive feedback reset with an in-pixel 1-bit image processing. In *Proc. SPIE*, volume 6068, pages 60680C-1-60680C-9, San Jose, January 2006.

[194] X. Wang, W. Wong, and R. Hornsey. A High Dynamic Range CMOS Image Sensor with Inpixel Light-to-Frequency Conversion. *IEEE Trans. Electron Devices*, 53(12): 2988-2992, December 2006.

[195] T. Serrano-Gotarredona, A. G. Andreou, and B. Linares-Barranco. AER image filtering architecture for vision-processing systems. *IEEE Trans. Circuits & Systems I*, 46(9): 1064-1071, September 1999.

[196] E. Culurciello, R. Etienne-Cummings, and K. A. Boahen. A Biomorphic Digital Image Sensor. *IEEE J. Solid-State Circuits*, 38(2): 281-294, February 2003.

[197] T. Teixeira, A. G. Andreou, and E. Culurciello. An Address-Event Image Sensor Network. In *Int'l Symp. Circuits & Systems (ISCAS)*, pages 644-647, Kobe, Japan, May 2005.

[198] T. Teixeira, E. Culurciello, and A. G. Andreou. An Address-Event Image Sensor Network. In *Int'l Symp. Circuits & Systems (ISCAS)*, pages 4467-4470, Kos, Greece, May 2006.

[199] M. L. Simpson, G. S. Sayler, G. Patterson, D. E. Nivens, E. K. Bolton, J. M. Rochelle, and J. C. Arnott. An integrated CMOS microluminometer for lowlevel luminescence sensing in the bioluminescent bioreporter integrated circuit. *Sensors & Actuators B*, 72: 134-140, 2001.

[200] E. K. Bolton, G. S. Sayler, D. E. Nivens, J. M. Rochelle, S. Ripp, and M. L. Simpson. Integratged CMOS photodetectors and signal processing for very low level chemical sensing with the bioluminescent bioreporter integrated circuits. *Sensors & Actuators B*, 85: 179-185, 2002.

[201] J. Ohta, N. Yoshida, K. Kagawa, and M. Nunoshita. Proposal of Application of Pulsed Vision Chip for Retinal Prosthesis. *Jpn. J. Appl. Phys.*, 41(4B): 2322-2325, April 2002.

[202] K. Kagawa, K. Isakari, T. Furumiya, A. Uehara, T. Tokuda, J. Ohta, and M. Nunoshita and. Pixel design of a pulsed CMOS image sensor for retinal prosthesis with digital photosensitivity control. *Electron. Lett.*, 39(5): 419-421, May 2003.

[203] A. Uehara, K. Kagawa, T. Tokuda, J. Ohta, and M. Nunoshita. Backilluminated pulse-frequency-modulated photosensor using a silicon-onsapphire technology developed for use as an epi-retinal prosthesis device. *Electron. Lett.*, 39(15): 1102-1104, July 2003.

[204] David C. Ng, K. Isakari, A. Uehara, K. Kagawa, T. Tokuda, J. Ohta, and M. Nunoshita. A study of bending effect on pulsed frequency modulation based photosensor for retinal prosthesis. *Jpn. J. Appl. Phys.*, 42(12): 7621-7624, December 2003.

[205] T. Furumiya, K. Kagawa, A. Uehara, T. Tokuda, J. Ohta, and M. Nunoshita. 32 * 32-pixel pulse-frequency-modulation based image sensor for retinal prosthesis. *J. Inst. Image Information & Television Eng.*, 58(3): 352-361, March 2004. In Japanese.

[206] K. Kagawa, N. Yoshida, T. Tokuda, J. Ohta, and M. Nunoshita. Building a Simple Model of A Pulse-Frequency-Modulation Photosensor and Demonstration of a 128 * 128-pixel Pulse-Frequency-ModulationImage Sensor Fabricated in a Standard 0.35-m Complementary Metal-Oxide Semiconductor Technology. *Opt. Rev.*, 11(3): 176-181, May 2004.

[207] A. Uehara, K. Kagawa, T. Tokuda, J. Ohta, and M. Nunoshita. A highsensitive digital photosensor using MOS interface-trap charge pumping. *IEICE Electronics Express*, 1(18): 556-561, December 2004.

[208] T. Furumiya, D. C. Ng, K. Yasuoka, K. Kagawa, T. Tokuda, M. Nunoshita, and J. Ohta. Functional verification of pulse frequency modulation-based image sensor for retinal prosthesis by *in vitro* electrophysiological experiments using frog retina. *Biosensors & Bioelectron.*, 21(7): 1059-1068, January 2006.

[209] S. Yamamoto, K. Kagawa, T. Furumiya, T. Tokuda, M. Nunoshita, and J. Ohta. Prototyping and evaluation of a 32 * 32-pixel pulse-frequency-modulation vision chip with capacitive-feedback reset. *J. Inst. Image Information & Television Eng.*, 60(4): 621-626, April 2006. In Japanese.

[210] T. Furumiya, S. Yamamoto, K. Kagawa, T. Tokuda, M. Nunoshita, and J. Ohta. Optimization of electrical stimulus pulse parameter for low-power operation of a retinal prosthetic device. *Jpn. J. Appl. Phys.*, 45(19): L505-L507, May 2006.

[211] D. C. Ng, T. Furumiya, K. Yasuoka, A. Uehara, K. Kagawa, T. Tokuda, M. Nunoshita, and J. Ohta. Pulse Frequency Modulation-based CMOS Image Sensor for Subretinal Stimulation. *IEEE Trans. Circuits & Systems II*, 53(6): 487-491, June 2006.

[212] J. Ohta, T. Tokuda, K. Kagawa, T. Furumiya, A. Uehara, Y. Terasawa, M. Oza-wa, T. Fujikado, and Y. Tano. Silicon LSI-Based Smart Stimulators for Retinal Prosthesis. *IEEE Eng. Medicine & Biology Magazine*, 25(5): 47-59, October 2006.

[213] J. Deguchi, T. Watanabe, T. Nakamura, Y. Nakagawa, T. Fukushima, S. Jeoung-Chill, H. Kurino, T. Abe, M. Tamai, and M. Koyanagi. Three Dimensionally Stacked Analog Retinal Prosthesis Chip. *Jpn. J. Appl. Phys.*, 43(4B): 1685-1689, April 2004.

[214] D. Ziegler, P. Linderholm, M. Mazza, S. Ferazzutti, D. Bertrand, A. M. Ionescu, and Ph. Renaud. An active microphotodiode array of oscillating pixels forretinal stimulation. *Sensors & Actuators A*, 110: 11-17, 2004.

[215] M. Mazza, P. Renaud, D. C. Bertrand, and A. M. Ionescu. CMOS Pixels for Subretinal Implantable Prothesis. *IEEE Sensors Journal*, 5(1): 32-27, Febru-ary 2005.

[216] M. L. Prydderch, M. J. French, K. Mathieson, C. Adams, D. Gunning, J. Laudanski, J. D. Morrison, A. R. Moodie, and J. Sinclair. A CMOS Active Pixel Sensor for Retinal Stimulation. In *Proc. SPIE*, pages 606803-1-606803-9, San Jose, 2006. Electronic Imaging.

[217] S. Kagami, T. Komuro, and M. Ishikawa. A Software-Controlled Pixel-Level ADC Conversion Method for Digital Vision Chips. In *IEEE Workshop on Charge-Coupled Devices & Advanced Image Sensors*, Elmau, Germany, May 2003.

[218] J. B. Kuoa and S. -C. Lin. *Low-voltage SOI CMOS VLSI devices and circuits*. John Wiley & Sons, Inc., New York, NY, 2001.

[219] A. Afzalian and D. Flandre. Modeling of the bulk versus SOI CMOS performances for the optimal design of APS circuits in low-power low-voltage applications. *IEEE Trans. Electron Devices*, 2003.

[220] K. Senda, E. Fujii, Y. Hiroshima, and T. Takamura. Smear-less SOI image sensor. In *Tech. Dig. Int'l Electron Devices Meeting (IEDM)*, pages 369-372, 1986.

[221] K. Kioi, T. Shinozaki, S. Toyoyama, K. Shirakawa, K. Ohtake, and S. Tsuchimoto. Design and implementation of a 3D-LSI image sensing processor. *IEEE J. Solid State Circuits*, 27(8): 1130-1140, August 1992.

[222] V. Suntharalingam, R. Berger, J. A. Burns, C. K. Chen, C. L. Keast, J. M. Knecht, R. D. Lambert, K. L. Newcomb, D. M. O'Mara, D. D. Rathman, D. C. Shaver, A. M. Soares, C. N. Stevenson, B. M. Tyrrell, K. Warner, B. D. Wheeler, D. -R. W. Yost, and D. J. Young. Megapixel CMOS image sensor fabricated in three-dimensional integrated circuit technology. In *Dig. Tech. Papers Int'l Solid-State Circuits Conf. (ISSCC)*, pages 356-357, February 2005.

[223] X. Zheng, S. Seshadri, M. Wood, C. Wrigley, and B. Pain. Process and Pixels for High Performance Imager in SOI-CMOS Technology. In *IEEE Workshop on Charge-Coupled Devices & Advanced Image Sensors*, Elmau, Germany, May 2003.

[224] B. Pain. Fabrication and Initial Results for a Back-illuminated Monolithic APS in a Mixed SOI/Bulk CMOS Technology. In *IEEE Workshop on Charge-Coupled Devices & Advanced Image Sensors*, pages 102-104, Karuizawa, Japan, June 2005.

[225] Y. S. Cho, H. Takano, K. Sawada, M. Ishida, and S. Y. Choi. SOI CMOS Image Sensor with Pinned Photodiode on Handle Wafer. In *IEEE Workshop on Charge-Coupled Devices & Advanced Image Sensors*, pages 105-108, Karuizawa, Japan, June 2005.

[226] S. Iwabuchi, Y. Maruyama, Y. Ohgishi, M. Muramatsu, N. Karasawa, and T. Hirayama. A Back-Illuminated High-Sensitivity Small-Pixel Color CMOS Image Sensor with Flexible Layout of Metal Wiring. In *Dig. Tech. Papers Int'l Solid-State Circuits Conf. (ISSCC)*, pages 1171-1178, February 2006.

[227] H. Yamamoto, K. Taniguchi, and C. Hamaguchi. High-sensitivity SOI MOS photodetectorwith self-amplification. *Jpn. J. Appl. Phys.*, 35(2B): 1382-1386, February 1996.

[228] W. Zhang, M. Chan, S. K. H. Fung, and P. K. Ko. Performance of a CMOS compatible lateral bipolar photodetector on SOI substrate. *IEEE Electron Device Lett.*, 19(11): 435-437, November 1998.

[229] W. Zhang, M. Chan, and P. K. Ko. Performance of the floating gate/body tied NMOSFET photodetector on SOI substrate. *IEEE Trans. Electron Devices*, 47(7): 1375-1384, July 2000.

[230] C. Xu, W. Zhang, and M. Chan. A low voltage hybrid bulk/SOI CMOS active pixel image sensor. *IEEE Electron Device Lett.*, 22(5): 248-250, May 2001.

[231] C. Xu, C. Shen, W. Wu, and M. Chan. Backside-Illuminated Lateral PIN Photodiode for CMOS Image Sensor on SOS Substrate. *IEEE Trans. Electron Devices*, 52(6): 1110-1115, June 2005.

[232] T. Ishikawa, M. Ueno, Y. Nakaki, K. Endo, Y. Ohta, J. Nakanishi, Y. Kosasayama, H. Yagi, T. Sone, and M. Kimata. Performance of 320 * 240 Uncooled IRFPA with SOI Diode Detectors. In *Proc. SPIE*, volume 4130, pages 152-159, 2000.

[233] A. G. Andreou, Z. K. Kalayjian, A. Apsel, P. O. Pouliquen, R. A. Athale, G. Simonis, and R. Reedy. Silicon on sapphire CMOS for optoelectronicmicrosystems. *IEEE Circuits & Systems Magazine*, 2001.

[234] E. Culurciello and A. G. Andreou. 16 * 16 pixel silicon on sapphire CMOS digital pixel photosensor array. *Electron. Lett.*, 40(1): 66-68, January 2004.

[235] A. Fish, E. Avner, and O. Yadid-Pecht. Low-power global/rolling shutter image sensors in silicon on sapphire technology. In *Int'l Symp. Circuits & Systems (ISCAS)*, pages 580-583, Kobe, Japan, May 2005.

[236] S. D. Gunapala, S. V. Bandara, J. K. Liu, Sir B. Rafol, and J. M. Mumolo. 640 * 512 Pixel Long-Wavelength Infrared Narrowband, Multiband, and Broadband QWIP Focal Plane Arrays. *IEEE Trans. Electron Devices*, 50(12): 2353-2360, December 2003.

[237] M. Kimata. Infrared Focal Plane Arrays. In H. Baltes, W. Gopel, and J. Hesse, editors, *Sensors Update*, volume 4, pages 53-79. Wiley-VCH, 1998.

[238] C.-C. Hsieh, C.-Y. Wu, F.-W. Jih, and T.-P. Sun. Focal-Plane-Arrays and CMOS Readout Techniques of Infrared Imaging Systems. *IEEE Trans. Circuits & Systems Video Tech.*, 7(4): 594-605, August 1997.

[239] M. Kimata. Silicon infrared focal plane arrays. In M. Henini and M. Razeghi, editors, *Handbook of Infrared Detection Technologies*, pages 352-392. Elsevier Science Ltd., 2002.

[240] E. Kasper and K. Lyutovich, editors. *Properties of Silicon Germanium and SiGe: Carbon*. INSPEC, The Institute of Electrical Engineers, London, UK, 2000.

[241] T. Tokuda, Y. Sakano, K. Kagawa, J. Ohta, and M. Nunoshita. Backsidehybirid photodetector for trans-chip detection of NIR light. In *IEEE Workshop on Charge-Coupled Devices & Advanced Image Sensors*, Elmau, Germany, May 2003.

[242] T. Tokuda, D. Mori, K. Kagawa, M. Nunoshita, and J. Ohta. A CMOS image sensor with eye-safe detection function using backside carrier injection. *J. Inst. Image Information & Television Eng.*, 60(3): 366-372, March 2006. In Japanese.

[243] J. A. Burns, B. F. Aull, C. K. Chen, C.-L. Chen, C. L. Keast, J. M. Knecht, V. Suntharalingam, K. Warner, P. W. Wyatt, and D.-R. W. Yost. A Wafer-Scale 3-D Circuit Integration Technology. *IEEE Trans. Electron Devices*, 53(10): 2507-2516, October 2006.

[244] M. Koyanagi, T. Nakamura, Y. Yamada, H. Kikuchi, T. Fukushima, T. Tanaka, and H. Kurino. Three-Dimensional Integration Technology Based on Wafer Bonding with Vertical Buried Interconnections. *IEEE Trans. Electron Devices*, 53(11): 2799-2808, November 2006.

[245] A. Iwata, M. Sasaki, T. Kikkawa, S. Kameda, H. Ando, K. Kimoto, D. Arizono, and H. Sunami. A 3D integration scheme utilizing wireless interconnections for implementing hyper brains. In *Dig. Tech. Papers Int'l Solid-State Circuits Conf. (ISSCC)*, pages 262-597, February 2005.

[246] N. Miura, D. Mizoguchi, M. Inoue, K. Niitsu, Y. Nakagawa, M. Tago, M. Fukaishi, T. Sakurai, and

T. Kuroda. A 1Tb/s 3W inductive-coupling transceiver for inter-chip clock and data link. In *Dig. Tech. Papers Int'l Solid-State Circuits Conf.* (*ISSCC*), pages 1676-1685, February 2006.

[247] E. Culurciello and A. G. Andreou. Capacitive Coupling of Data and Power for 3D Silicon-on-Insulator VLSI. In *Int'l Symp. Circuits & Systems* (*ISCAS*), pages 4142-4145, Kobe, Japan, May 2005.

[248] D. A. B. Miller, A. Bhatnagar, S. Palermo, A. Emami-Neyestanak, and M. A. Horowitz. Opportunities for Optics in Integrated Circuits Applications. In *Dig. Tech. Papers Int'l Solid-State Circuits Conf.* (*ISSCC*), pages 86-87, February 2005.

[249] K. W. Lee, T. Nakamura, K. Sakuma, K. T. Park, H. Shimazutsu, N. Miyakawa, K. Y. Kim, H. Kurino, and M. Koyanagi. Development of Three-Dimensional Integration Technology for Highly Parallel Image-Processing Chip. *Jpn. J. Appl. Phys.*, 39(4B): 2473-2477, April 2000.

[250] M. Koyanagi, Y. Nakagawa, K. W. Lee, T. Nakamura, Y. Yamada, K. Inamura, K.-T. Park, and H. Kurino. Neuromorphic Vision Chip Fabricated Using Three-Dimensional Integration Technology. In *Dig. Tech. Papers Int'l Solid-State Circuits Conf.* (*ISSCC*), February 2001.

[251] S. Lombardo, S. U. Campisano, G. N. van den Hoven, and A. Polman. Erbium in oxygen-doped silicon: Electroluminescence. *J. Appl. Phys.*, 77(12): 6504-6510, December 1995.

[252] M. A. Green, J. Zhao, A. Wang, P. J. Reece, and M. Gal. Efficient silicon light-emitting diodes. *Nature*, 412: 805-808, 2001.

[253] L. W. Snyman, M. du Plessis, E. Seevinck, and H. Aharoni. An Efficient Low Voltage, High Frequency Silicon CMOS Light Emitting Device and Electro-Optical Interface. *IEEE Trans. Electron Devices*, 20(12): 614-617, December 1999.

[254] M. du Plessis, H. Aharoni, and L. W. Snyman. Spatial and intensity modulation of light emission from a silicon LED matrix. *IEEE Photon. Tech. Lett.*, 14(6): 768-770, June 2002.

[255] J. Ohta, K. Isakari, H. Nakayama, K. Kagawa, T. Tokuda, and M. Nunoshita. An image sensor integrated with light emitter using BiCMOS process. *J. Inst. Image Information & Television Eng.*, 57(3): 378-383, March 2003. In Japanese.

[256] H. Aharoni and M. du Plessis. Low-operating-voltage integrated silicon lightemitting devices. *IEEE J. Quantum Electron.*, 40(5): 557-563, May 2004.

[257] M. Sergio and E. Charbon. An intra-chip electro-optical channel based on CMOS single photon detectors. In *Tech. Dig. Int'l Electron Devices Meeting* (*IEDM*), December 2005.

[258] J. R. Haynes and W. C. Westphal. Radiation Resulting from Recombination of Holes and Electrons in Silicon. *Phys. Rev.*, 101(6): 1676-1678, March 1956.

[259] S. Maëstre and P. Magnan. Electroluminescence and Impact Ionization in CMOS Active Pixel Sensors. In *IEEE Workshop on Charge-Coupled Devices & Advanced Image Sensors*, Elmau, Germany, May 2003.

[260] K. M. Findlater, D. Renshaw, J. E. D. Hurwitz, R. K. Henderson, T. E. R. Biley, S. G. Smith, M. D. Purcell, and J. M. Raynor. A CMOS Image Sensor Employing a Double Junction Photodiode. In *IEEE Workshop on Charge-Coupled Devices & Advanced Image Sensors*, pages 60-63, Lake Tahoe, NV, June 2001.

[261] D. L. Gilblom, S. K. Yoo, and P. Ventura. Operation and performance of a color image sensor with layered photodiodes. In *Proc. SPIE*, volume 5074 of *SPIE AeroSense*, pages 5210-14-27, Orlando, FL, April 2003.

[262] R. B. Merrill. Color Separation in an Active Pixel Cell Imaging Array Using a Triple-Well Structure. US Patent 5,965,875, 1999.

[263] T. Lulé, B. Schneider, and M. Böhm. Design and Fabrication of a High-Dynamic-Range Image Sensor in TFA Technology. *IEEE J. Solid-State Circuits*, 34(5): 704-711, May 1999.

[264] M. Sommer, P. Rieve, M. Verhoeven, M. Böhm, B. Schneider, B. van Uffel, and F. Librecht. First Multispectral Diode Color Imager with Three Color Recognition and Color Memory in Each Pixel. In *IEEE Workshop on Charge-Coupled Devices & Advanced Image Sensors*, pages 187-190, Karuizawa, Japan, June 1999.

[265] H. Steibig, R. A. Street, D. Knipp, M. Krause, and J. Ho. Vertically integrated thin-film color sensor arrays for advanced sensing applications. *Appl. Phys. Lett.*, 88: 013509, 2006.

[266] Y. Maruyama, K. Sawada, H. Takao, and M. Ishida. The fabrication of filterless luorescence detection sensor array using CMOS image sensor technique. *ensors & Actuators A*, 128: 66-70, 2006.

[267] Y. Maruyama, K. Sawada, H. Takao, and M. Ishida. A novel filterless fluorescence detection sensor for DNA analysis. *IEEE Trans. Electron Devices*, 53(3): 553-558, March 2006.

[268] P. Catrysse, B. Wandell, and A. El Gamal. An integrated color pixel in 0.18μm CMOS technology. In *Tech. Dig. Int'l Electron Devices Meeting (IEDM)*, pages 24.4.1-24.4.4, December 2001.

[269] K. Sasagawa, K. Kusawake, K. Kagawa, J. Ohta, and M. Nunoshita. Optical transmission enhancement for an image sensor with a sub-wavelength aperture. In *Int'l Conf. Optics-Photonics Design & Fabrication (ODF)*, pages 163-164, November 2002.

[270] L. W. Barnes, A. Dereux, and T. W. Ebbesen. Surface plasmon subwavelength optics. *Nature*, 2003.

[271] Y. Inaba, M. Kasano, K. Tanaka, and T. Yamaguchi. Degradation-free MOS image sensor with photonic crystal color filter. *IEEE Electron Device Lett.*, 27(6): 457-459, June 2006.

[272] H. A. Bethe. Theory of Diffraction by Small Holes. *Phys. Rev.*, 66(7-8): 163-182, October 1944.

[273] T. Thio, K. M. Pellerin, R. A. Linke, H. J. Lezec, and T. W. Ebbessen. Enhanced light transmission through a single subwavelength aperture. *Opt. Lett.*, 26, December 2001.

[274] H. Raether. *Surface Plasmons on Smooth and Rough Surfaces and on Gratings*. Springer-Verlag, Berlin, Germany, 1988.

[275] B. J. Hosticka, W. Brockherde, A. Bussmann, T. Heimann, R. Jeremias, A. Kemna, C. Nitta, and O. Schrey. CMOS imaging for automotive applications. *IEEE Trans. Electron Devices*, 50(1): 173-183, January 2003.

[276] J. Hynecek. Impactron—A New Solid State Image Intensifier. In *IEEE Workshop on Charge-Coupled Devices & Advanced Image Sensors*, pages 197-200, Lake Tahoe, NV, June 2001.

[277] S. Ohta, H. Shibuya, I. Kobayashi, T. Tachibana, T. Nishiwaki, and J. Hynecek. Characterization Results of 1k * 1k Charge Multiplying CCD Image Sensor. In *Proc. SPIE*, volume 5301, pages 99-108, San Jose, CA, January 2004.

[278] H. Eltoukhy, K. Salama, and A. E. Gamal. A 0.18-μm CMOS bioluminescence detection lab-on-chip. *IEEE J. Solid-State Circuits*, 2006.

[279] H. Jia and P. A. Abshire. A CMOS image sensor for low light applications. In *Int'l Symp. Circuits & Systems (ISCAS)*, pages 1651-1654, Kos, Greece, May 2006.

[280] B. Fowler, M. D. Godfrey, J. Balicki, and J. Canfield. Low Noise Readout using Active Reset for CMOS APS. In *Proc. SPIE*, volume 3965, pages 126-135, San Jose, CA, January 2000.

[281] I. Takayanagi, Y. Fukunaga, T. Yoshida, and J. Nakamura. A Four-Transistor Capacitive Feedback Reset Active Pixel and its Reset Noise Reduction Capability. In *IEEE Workshop on Charge-Coupled Devices & Advanced Image Sensors*, pages 118-121, Lake Tahoe, NV, June 2001.

[282] B. Pain, T. J. Cunningham, B. Hancock, G. Yang, S. Seshadri, and M. Ortiz. Reset noise suppression in two-dimensional CMOS photodiode pixels through column-based feedback-reset. In *Tech. Dig. Int'l Electron Devices Meeting (IEDM)*, pages 809-812, December 2002.

[283] S. Kleinfelder. High-speed, high-sensitivity, low-noise CMOS scientific image sensors. In *Proc. SPIE*, volume 5247 of *Microelectronics: Design, Technology, and Packaging*, pages 194-205,

December 2003.

[284] T. J. Cunningham, B. Hancock, C. Sun, G. Yang, M. Oritz, C. Wrigley, S. Seshadri, and B. Pain. A Two-Dimensional Array Imager Demonstrting Active Reset Suppression of kTC-Noise. In *IEEE Workshop on Charge-Coupled Devices & Advanced Image Sensors*, Elmau, Germany, May 2003.

[285] Y. Chen and S. Kleinfelder. CMOS active pixel sensor achieving 90dB dynamic range with colum-level active reset. In *Proc. SPIE*, volume 5301, pages 438-449, San Jose, CA, 2004.

[286] K.-H. Lee and E. Yoon. A CMOS Image Sensor with Reset Level Control Using Dynamc Reset Current Source for Noise Suppression. In *Dig. Tech. Papers Int'l Solid-State Circuits Conf. (ISSCC)*, page 114, February 2004.

[287] L. Kozlowski, G. Rossi, L. Blanquart, R. Marchesini, Y. Huang, G. Chow, and J. Richardson. A Progressive 1920 * 1080 Imaging System-on-Chip for HDTV Cameras. In *Dig. Tech. Papers Int'l Solid-State Circuits Conf. (ISSCC)*, pages 358-359, February 2005.

[288] J. Yang, K. G. Fife, L. Brooks, C. G. Sodini, A. Betts, P. Mudunuru, and H.-S. Lee. A 3M Pixel Low-Noise Flexible Architecture CMOS Image Sensor. In *Dig. Tech. Papers Int'l Solid-State Circuits Conf. (ISSCC)*, pages 2004-2013, February 2006.

[289] T. G. Etoh, D. Poggemann, G. Kreider, H. Mutoh, A. J. P. Theuwissen, A. Ruckelshausen, Y. Kondo, H. Maruno, K. Takubo, H. Soya, K. Takehara, T. Okinaka, and Y. Takano. An image sensor which captures 100 consecutive frames at 1000000 frames/s. *IEEE Trans. Electron Devices*, 50(1): 144-151, January 2003.

[290] A. I. Krymski, N. E. Bock, N. Tu, D. Van Blerkom, and E. R. Fossum. A High-Speed, 240-Frames/s, 4. 1-Mpixel CMOS Sensor. *IEEE Trans. Electron Devices*, 50(1): 130-135, January 2003.

[291] A. I. Krymski and N. Tu. A 9-V/Lux-s 5000-frames/s 512 * 512 CMOS sensor. *IEEE Trans. Electron Devices*, 50(1): 136-143, January 2003.

[292] Y. Nitta, Y. Muramatsu, K. Amano, T. Toyama, J. Yamamoto, K. Mishina, A. Suzuki, T. Taura, A. Kato, M. Kikuchi, Y. Yasui, H. Nomura, and N. Fukushima. High-Speed Digital Double Sampling with Analog CDS on Column Parallel ADC Architecture for Low-Noise Active Pixel Sensor. In *Dig. Tech. Papers Int'l Solid-State Circuits Conf. (ISSCC)*, pages 2024-2031, February 2006.

[293] A. I. Krymski and K. Tajima. CMOS Image Sensor with integrated 4Gbps Camera Link Transmitter. In *Dig. Tech. Papers Int'l Solid-State Circuits Conf. (ISSCC)*, pages 2040-2049, February 2006.

[294] M. Furuta, T. Inoue, Y. Nishikawa, and S. Kawahito. A 3500fps High-Speed CMOS Image Sensor with 12b Column-Parallel Cyclic A/D Converters. In *Dig. Tech. Papers Symp. VLSI Circuits*, pages 21-22, June 2006.

[295] T. Inoue, S. Takeuchi, and S. Kawhito. CMOS active pixel image sensor with in-pixel CDS for high-speed cameras. In *Proc. SPIE*, volume 5301, pages 2510-257, San Jose, CA, January 2004.

[296] S. Lauxtermann, G. Israel, P. Seitz, H. Bloss, J. Ernst, H. Firla, and S. Gick. A mega-pixel high speed CMOS sensor with sustainable Gigapixel/s readout rate. In *IEEE Workshop on Charge-Coupled Devices & Advanced Image Sensors*, pages 48-51, Lake Tahoe, NV, June 2001.

[297] N. Stevanovic, M. Hillebrand, B. J. Hosticka, and A. Teuner. A CMOS image sensor for high-speed imaging. In *Dig. Tech. Papers Int'l Solid-State Circuits Conf. (ISSCC)*, pages 104-105, February 2000.

[298] N. Bock, A. Krymski, A. Sarwari, M. Sutanu, N. Tu, K. Hunt, M. Cleary, N. Khaliullin, and M. Brading. A wide-VGA CMOS image sensor with global shutter and extended dyamic range. In *IEEE Workshop on Charge-CoupledDevices & Advanced Image Sensors*, pages 222-225, Karuizawa, Japan, June 2005.

[299] G. Yang and T. Dosluoglu. Ultra High Light Shutter Rejection Ratio Snapshot Pixel Image Sensor

ASIC for Pattern Recoginition. In *IEEE Workshop on Charge-Coupled Devices & Advanced Image Sensors*, pages 161-164, Karuizawa, Japan, June 2005.

[300] D. H. Hubel. *Eye, Brain, and Vision*. Scientific American Library, New York, NY, 1987.

[301] M. Mase, S. Kawahito, M. Sasaki, Y. Wakamori, and M. Furuta. A Wide Dynamic Range CMOS Image Sensor with Multiple Exposure-Time Signal Outputs and 12-bit Column-Parallel Cyclic AD Converters. *IEEE J. Solid-State Circuits*, 40(12): 2787-2795, December 2005.

[302] Y. Wang, S. L. Barna, S. Campbell, and E. R. Fossum. A high dynamic range CMOS APS image sensor. In *IEEE Workshop on Charge-Coupled Devices & Advanced Image Sensors*, pages 137-140, Lake Tahoe, NV, June 2001.

[303] D. Yang and A. El Gamal. Comparative analysis of SNR for image sensors with widened dynamic range. In *Proc. SPIE*, volume 3649, pages 197-211, San Jose, CA, February 1999.

[304] R. Hauschild, M. Hillebrand, B. J. Hosticka, J. Huppertz, T. Kneip, and M. Schwarz. A CMOS image sensor with local brightness adaptation and high intrascene dynamic range. In *Proc. European Solid-State Circuits Conf. (ESSCIRC)*, pages 308-311, September 1998.

[305] O. Schrey, R. Hauschild, B. J. Hosticka, U. Iurgel, and M. Schwarz. A locally adaptive CMOS image sensor with 90dB dynamic range. In *Dig. Tech. Papers Int'l Solid-State Circuits Conf. (ISSCC)*, pages 310-311, February 1999.

[306] S. L. Barna, L. P. Ang, B. Mansoorian, and E. R. Fossum. A low-light to sunlight, 60 frames/s, 80k pixel CMOS APS camera-on-a-chip with 8b digital output. In *IEEE Workshop on Charge-Coupled Devices & Advanced Image Sensors*, pages 148-150, Karuizawa, Japan, June 1999.

[307] S. Lee and K. Yang. High dynamic-range CMOS image sensor cell based on self-adaptive photosensing operation. *IEEE Trans. Electron Devices*, 53(7): 1733-1735, July 2006.

[308] S. Chamberlain and J. P. Y. Lee. A novel wide dynamic range silicon photodetector and liner imaging array. *IEEE J. Solid-State Circuits*, 19(1), February 1984.

[309] G. G. Storm, J. E. D. Hurwitz, D. Renshawa, K. M. FIndlater, R. K. Henderson, and M. D. Purcell. High dynamic range imaging using combined linearlogarithmic response from a CMOS image sensor. In *IEEE Workshop on Charge-Coupled Devices & Advanced Image Sensors*, Elmau, Germany, May 2003.

[310] K. Hara, H. Kubo, M. Kimura, F. Murao, and S. Komori. A Linear-Logarithmic CMOS Sensor with Offset Calibration Using an Injected Charge Signal. In *Dig. Tech. Papers Int'l Solid-State Circuits Conf. (ISSCC)*, 2005.

[311] S. Decker, D. McGrath, K. Brehmer, and C. G. Sodini. A 256 * 256 CMOS imaging array with wide dynamic range pixels and column-parallel digital output. *IEEE J. Solid-State Circuits*, 33(12): 2081-2091, December 1998.

[312] Y. Muramatsu, S. Kurosawa, M. Furumiya, H. Ohkubo, and Y. Nakashiba. A Signal-Processing CMOS Image Sensor using Simple Analog Operation. In *Dig. Tech. Papers Int'l Solid-State Circuits Conf. (ISSCC)*, pages 98-99, February 2001.

[313] V. Berezin, I. Ovsiannikov, D. Jerdev, and R. Tsai. Dynamic Range Enlargement in CMOS Imagers with Buried Photodiode. In *IEEE Workshop on Charge-Coupled Devices & Advanced Image Sensors*, Elmau, Germany, 2003.

[314] S. Sugawa, N. Akahane, S. Adachi, K. Mori, T. Ishiuchi, and K. Mizobuchi. A 100dB dyamic range CMOS image sensor using a lateral overflowintegration capacitor. In *Dig. Tech. Papers Int'l Solid-State Circuits Conf. (ISSCC)*, pages 352-353, February 2005.

[315] N. Akahane, R. Ryuzaki, S. Adachi, K. Mizobuchi, and S. Sugawa. A 200dB Dynamic Range Iris-less CMOS Image Sensor with Lateral Overflow Integration Capacitor using Hybrid Voltage and Current

Readout Operation. In *Dig. Tech. Papers Int'l Solid State Circuits Conf.* (ISSCC), February 2006.

[316] N. Akahane, S. Sugawa, S. Adachi, K. Mori, T. Ishiuchi, and K. Mizobuchi. A sensitivity and linearity improvement of a 100-dB dynamic range CMOS image sensor using a lateral overflow integration capacitor. *IEEE J. Solid-State Circuits*, 41(4): 851-858, April 2006.

[317] M. Ikeba and K. Saito. CMOS-Image Sensor with PD-Capacitance Modulation using Negative Feedback Resetting. *J. Inst. Image Information & Television Eng.*, 60(3): 384, March 2006.

[318] K. Kagawa, Y. Adachi, Y. Nose, H. Takashima, K. Tani, A. Wada, M. Nunoshita, and J. Ohta. A wide-dynamic-range CMOS imager by hybrid use of active and passive pixel sensors. In *IS&T SPIE Annual Symposium Electronic Imaging*, [6501-18], San Jose, CA, January 2007.

[319] S. T. Smith, P. Zalud, J. Kalinowski, N. J. McCaffrey, P. A. Levine, and M. L. Lin. BLINC: a 640 * 480 CMOS active pixel vide camera with adaptive digital processing, extended optical dynamic range, and miniature form factor. In *Proc. SPIE*, volume 4306, pages 41-49, January 2001.

[320] H. Witter, T. Walschap, G. Vanstraelen, G. Chapinal, G. Meynants, and B. Dierickx. 1024 * 1280 pixel dual shutter APS for industrial vision. In *Proc. SPIE*, volume 5017, pages 19-23, 2003.

[321] O. Yadid-Pecht and E. R. Fossum. Wide intrascene dynamic range CMOS APS using dual sampling. *IEEE Trans. Electron Devices*, 44(10): 1721-1723, October 1997.

[322] O. Schrey, J. Huppertz, G. Filimonovic, A. Bussmann, W. Brockherde, and B. J. Hosticka. A1 K * 1 K high dynamic range CMOS image sensor with on-chip programmable region-of-interest. *IEEE J. Solid-State Circuits*, 37(7): 911-915, September 2002.

[323] K. Mabuchi, N. Nakamura, E. Funatsu, T. Abe, T. Umeda, T. Hoshino, R. Suzuki, and H. Sumi. CMOS image sensor using a floating diffusion driving buried photodiode. In *Dig. Tech. Papers Int'l Solid-State Circuits Conf.* (ISSCC), pages 112-516, February 2004.

[324] M. Sasaki, M. Mase, S. Kawahito, and Y. Tadokoro. A Wide Dyamic Range CMOS Image Sensor wit Integration of Short-Exposure-Time Signals. In *IEEE Workshop on Charge-Coupled Devices & Advanced Image Sensors*, Elmau, Germany, 2003.

[325] M. Sasaki, M. Mase, S. Kawahito, and Y. Tadokoro. A wide dynamic range CMOS image sensor with multiple short-time exposures. In *Proc. IEEE Sensors*, volume 2, pages 967-972, October 2004.

[326] M. Schanz, C. Nitta, A. Bußmann, B. J. Hosticka, and R. K. Wertheimer. A high-dynamic-range CMOS image sensor for automotive applications. *IEEE J. Solid-State Circuits*, 35(7): 932-938, September 2000.

[327] B. Schneider, H. FIsher, S. Benthien, H. Keller, T. Lul'e, P. Rieve, M. SOmmer, and J. Schulte M. Böhm. TFA Image Sensors: From the One Transistor Cell to a Locally Adaptive High Dynamic Range Sensor. In *Tech. Dig. Int'l Electron Devices Meeting* (IEDM), pages 209-212, December 1997.

[328] T. Hamamoto and K. Aizawa. A computational image sensor with adaptive pixel-based integration time. *IEEE J. Solid-State Circuits*, 36(4): 580-585, April 2001.

[329] T. Yasuda, T. Hamamoto, and K. Aizawa. Adaptive-integration-time image sensor with real-time reconstruction function. *IEEE Trans. Electron Devices*, 50(1): 111-120, January 2003.

[330] O. Yadid-Pecht and A. Belenky. In-pixel autoexposure CMOS APS. *IEEE J. Solid-State Circuits*, 38(8): 1425-1428, August 2003.

[331] P. M. Acosta-Serafini, I. Masaki, and C. G. Sodini. A 1/3" VGA linear wide dynamic range CMOS image sensor implementing a predictive multiple sampling algorithm with overlapping integration intervals. *IEEE J. Solid-State Circuits*, 39(9): 1487-1496, September 2004.

[332] T. Anaxagoras and N. M. Allinson. High dynamic range active pixel sensor. In *Proc. SPIE*, volume 5301, pages 149-160, San Jose, CA, January 2004.

[333] D. Stoppa, A. Simoni, L. Gonzo, M. Gottardi, and G. -F. Dalla Betta. Novel CMOS image sensor with a 132-dB dynamic range. *IEEE J. Solid-State Circuits*, 37(12): 1846-1852, December 2002.

[334] J. Döge, G. Shöneberg, G. T. Streil, and A. König. An HDR CMOS Image Sensor with Spiking Pixels, Pixel-Level ADC, and Linear Characteristics. *IEEE Trans. Circuits & Systems II*, 49(2): 155-158, February 2002.

[335] S. Chen, A. Bermak, and F. Boussaid. A compact reconfigurable counter memory for spiking pixels. *IEEE Electron Device Lett.*, 27(4): 255-257, April 2006.

[336] K. Oda, H. Kobayashi, K. Takemura, Y. Takeuchi, and T. Yamda. The development of wide dynamic range iamge sensor. *ITE Tech. Report*, 27(25): 17-20, 2003. In Japanese.

[337] S. Ando and A. Kimachi. Time-Domain Correlation Image Sensor: First CMOS Realization of Demodulation Pixels Array. In *IEEE Workshop on Charge-Coupled Devices & Advanced Image Sensors*, pages 33-36, Karuizawa, Japan, June 1999.

[338] T. Spirig, P. Seitz, O. Vietze, and F. Heitger. The lock-in CCD-twodimensional synchronous detection of light. *IEEE J. Quantum Electron*, 31(9): 1705-1708, September 1995.

[339] T. Spirig, M. Marley, and P. Seitz. The multitap lock-in CCD with offset subtraction. *IEEE Trans. Electron Devices*, 44(10): 1643-1647, October 1997.

[340] R. Lange, P. Seitz, A. Biber, and S. Lauxtermann. Demodulation Pixels in CCD and CMOS Technologies for Time-Of-Flight Ranging. In *Proc. SPIE*, pages 177-188, San Jose, CA, January 2000.

[341] R. Lange and P. Seitz. Solid-state time-of-flight range camera. *IEEE J. Quantum Electron*, 37(3): 390-397, March 2001.

[342] A. Kimachi, T. Kurihara, M. Takamoto, and S. Ando. A Novel Range Finding Systems Using Correlation Image Sensor. *Trans. IEE Jpn.*, 121-E(7): 367-375, July 2001.

[343] K. Yamamoto, Y. Oya, K. Kagawa, J. Ohta, M. Nunoshita, and K. Watanabe. Demonstration of a freqency-demodulation CMOS image sensor and its improvement of image quality. In *IEEE Workshop on Charge-Coupled Devices & Advanced Image Sensors*, Elmau, Germany, June 2003.

[344] S. Ando and A. Kimachi. Correlation Image Sensor: Two-Dimensional Matched Detection of Amplitude-Modulated Light. *IEEE Trans. Electron Devices*, 50(10): 2059-2066, October 2003.

[345] R. Miyagawa and T. Kanade. CCD-based range-finding sensor. *IEEE Trans. Electron Devices*, 44(10): 1648-1652, October 1997.

[346] B. Buxbaum, R. Chwarte, and T. Ringbeck. PMD-PLL: Receiver structure for incoherent communication and ranging systems. In *Proc. SPIE*, volume 3850, pages 116-127, 1999.

[347] S. -Y. Ma and L. -G. Chen. A single-chip CMOS APS camera with direct frame difference output. *IEEE J. Solid-State Circuits*, 34(10): 1415-1418, October 1999.

[348] J. Ohta, K. Yamamoto, T. Hirai, K. Kagawa, M. Nunoshita, M. Yamada, Y. Yamasaki, S. Sughishita, and K. Watanabe. An image sensor with an inpixel demodulation function for detecting the intensity of a modulated light signal. *IEEE Trans. Electron Devices*, 50(1): 166-172, January 2003.

[349] K. Yamamoto, Y. Oya, K. Kagawa, J. Ohta, M. Nunoshita, and K. Watanabe. Improvement of demodulated image quality in a demodulated image sensor. *J. Inst. Image Information & Television Eng.*, 57(9): 1108-1114, September 2003. In Japanese.

[350] Y. Oike, M. Ikeda, and K. Asada. A 120×110 position sensor with the capability of sensitive and selective light detection in wide dynamic range for robust active range finding. *IEEE J. Solid-State Circuits*, 39(1): 246-251, January 2004.

[351] Y. Oike, M. Ikeda, and K. Asada. Pixel-Level Color Demodulation Image Sensor for Support of Image Recognition. *IEICE Trans. Electron.*, E87-C(12): 2164-2171, December 2004.

[352] K. Yamamoto, Y. Oya, K. Kagawa, M. Nunoshita, J. Ohta, and K. Watanabe. A 128 × 128 Pixel CMOS Image Sensor with an Improved Pixel Architecture for Detecting Modulated Light Signals. *Opt. Rev.*, 13(2): 64-68, April 2006.

[353] B. Buttgen, F. Lustenberger, and P. Seitz. Demodulation Pixel Based on Static Drift Fields. *IEEE Trans. Electron Devices*, 53(11): 2741-2747, November 2006.

[354] A. Kimachi, H. Ikuta, Y. Fujiwara, and H. Matsuyama. Spectral matching imager using correlation image sensor and AM-coded multispectral illuminatin. In *Proc. SPIE*, volume 5017, pages 128-135, Santa Clara, CA, January 2003.

[355] K. Asashi, M. Takahashi, K. Yamamoto, K. Kagawa, and J. Ohta. Application of a demodulated image sensos for a camera sysytem suppressing saturation. *J. Inst. Image Information & Television Eng.*, 60(4): 627-630, April 2006. In Japanese.

[356] Y. Oike, M. Ikeda, and K. Asada. Design and implementation of real-time 3-D image sensor with 640×480 pixel resolution. *IEEE J. Solid-State Circuits*, 39(4): 622-628, April 2004.

[357] B. Aull, J. Burns, C. Chen, B. Felton, H. Hanson, C. Keast, J. Knecht, A. Loomis, M. Renzi, A. Soares, V. Suntharalingam, K. Warner, D. Wolfson, D. Yost, and D. Young. Laser Radar Imager Based on 3D Integration of Geiger-Mode Avalanche Photodiodes with Two SOI Timing Circuit Layers. In *Dig. Tech. Papers Int'l Solid-State Circuits Conf.* (ISSCC), pages 1179-1188, February 2006.

[358] L. Viarani, D. Stoppa, L. Gonzo, M. Gottardi, and A. Simoni. A CMOS Smart Pixel for Active 3-D Vision Applications. *IEEE Sensors Journal*, 4(1): 145-152, February 2004.

[359] D. Stoppa, L. Viarani, A. Simoni, L. Gonzo, M. Malfatti, and G. Pedretti. A 50 × 50-pixel CMOS sensor for TOF-based real time 3D imaging. In *IEEE Workshop on Charge-Coupled Devices & Advanced Image Sensors*, pages 230-233, Karuizawa, Japan, June 2005.

[360] R. Jeremias, W. Brockherde, G. Doemens, B. Hosticka, L. Listl, and P. Mengel. A CMOS photosensor array for 3D imaging using pulsed laser. In *Dig. Tech. Papers Int'l Solid-State Circuits Conf.* (ISSCC), pages 252-253, February 2001.

[361] O. Elkhalili, O. M. Schrey, P. Mengel, M. Petermann, W. Brockherde, and B. J. Hosticka. A 4 × 64 pixel CMOS image sensor for 3-D measurement applications. *IEEE J. Solid-State Circuits*, 30(7): 1208-1212, February 2004.

[362] T. Sawasa, T Ushinaga, I. A. Halin, S. Kawahito, M. Homma, and Y. Maeda. A QVGA-size CMOS time-of-flight range image sensor with background ligh charge draining structure. *ITE Tech. Report*, 30(25): 13-16, March 2006. In Japanese.

[363] T. Moller, H. Kraft, J. Frey, M. Albrecht, and R. Lange. Robust 3D Measurement with PMD Sensors. In *Proc. 1st Range Imaging Research Day at ETH Zurich*, page Supplement to the Proceedings, Zurich, 2005.

[364] T. Kahlmann, F. Remondino, and H. Ingensand. Calibration for increased accuracy of the range imaging camera Swiss Ranger TM. In *Int'l Arch. Photogrammetry, Remote Sensing & Spatial Information Sci.*, volume XXXVI part 5, pages 136-141. Int'l Soc. Photogrammetry & Remote Sensing (ISPRS) Commission V Symposium, September 2006.

[365] M. Lehmann, T. Oggier, B. Büttgen, Chr. Gimkiewicz, M. Schweizer, R. Kaufmann, F. Lustenberger, and N. Blanc. Smart pixels for future 3DTOF sensors. In *IEEE Workshop on Charge-Coupled Devices & Advanced Image Sensors*, pages 193-196, Karuizawa, Japan, June 2005.

[366] S. B. Gokturk, H. Yalcin, and C. Bamji. A Time-of-Flight Depth Sensor—System Description, Issues and Solutions. In *Conf. Computer Vision & Pattern Recognition Workshop* (CVPR), pages 35-44, Washington, DC, June 2004.

[367] B. Pain, L. Matthies, B. Hancock, and C. Sun. A compact snap-shot rangeimaging receiver. In *IEEE Workshop on Charge-Coupled Devices & Advanced Image Sensors*, pages 234-237, Karuizawa, Japan, June 2005.

[368] T. Kato, S. Kawahito, K. Kobayashi, H. Sasaki, T. Eki, and T. Hisanaga. A Bioncular CMOS Range Image Sensor with Bit-Serial Block-Parallel Interface Using Cyclic Pipelined ADC's. In *Dig. Tech. Papers Symp. VLSI Circuits*, pages 270-271, Honolulu, Hawaii, June 2002.

[369] S. Kakehi, S. Nagao, and T. Hamamoto. Smart Image Sensor with Binocular PD Array for Tracking of a Moving Object and Depth Estimation. In *Int'l Symposium Intelligent Signal Processing & Communication Systems (ISPACS)*, pages 635-638, Awaji, Japan, December 2003.

[370] R. M. Philipp and R. Etienne-Cummings. Single chip stero imager. In *Int'l Symp. Circuits & Systems (ISCAS)*, pages 808-811, 2003.

[371] R. M. Philipp and R. Etienne-Cummings. A 128×128 33mW 30frames/s single-chip stereo imager. In *Dig. Tech. Papers Int'l Solid-State Circuits Conf. (ISSCC)*, pages 2050-2059, February 2006.

[372] A. Gruss, L. R. Carley, and T. Kanade. Integrated Sensor and Range-Finding Analog Signal Processor. *IEEE J. Solid-State Circuits*, 26(3): 184-191, March 1991.

[373] Y. Oike, M. Ikeda, and K. Asada. A CMOS Image Sensor for High-Speed Active Range Finding Using Column-Parallel Time-Domain ADC and Position Encoder. *IEEE Trans. Electron Devices*, 50(1): 152-158, January 2003.

[374] Y. Oike, M. Ikeda, and K. Asada. A 375×365 high-speed 3-D range-finding image sensor using row-parallel search architecture and multisampling technique. *IEEE J. Solid-State Circuits*, 40(2): 444-453, February 2005.

[375] T. Sugiyama, S. Yoshimura, R. Suzuki, and H. Sumi. A 1/4-inch QVGA color imaging and 3-D sensing CMOS sensor with analog frame memory. In *Dig. Tech. Papers Int'l Solid-State Circuits Conf. (ISSCC)*, pages 434-479, February 2002.

[376] H. Miura, H. Ishiwata, Y. Lida, Y. Matunaga, S. Numazaki, A. Morisita, N. Umeki, and M. Doi. 100 frame/s CMOS active pixel sensor for 3Dgesture recognition system. In *Dig. Tech. Papers Int'l Solid-State Circuits Conf. (ISSCC)*, pages 142-143, February 1999.

[377] M. D. Adams. Coaxial Range Measurement—Current Trends for Mobile Robotic Applications. *IEEE Sensors Journal*, 2(1): 2-13, February 2002.

[378] D. Vallancourt and S. J. Daubert. Applications of current-copier circuits. In C. Toumazou, F. J. Lidgey, and D. G. Haigh, editors, *Analogue IC design: the current-mode approach*, IEEE Circuts and Systems Series 2, chapter 14, pages 515-533. Peter Peregrinus Ltd., London, UK, 1990.

[379] K. Aizawa, K. Sakaue, and Y. Suenaga, editors. *Image Processig Technologies, Alogorithms, Sensors, and Applications*. Mracel Dekker, Inc., New York, NY, 2004.

[380] A. Yokota, T. Yoshida, H. Kashiyama, and T. Hamamoto. High-speed Sensing System for Depth Estimation Based on Depth-from-Focus by Using Smart Imager. In *Int'l Symp. Circuits & Systems (ISCAS)*, pages 564-567, Kobe, Japan, May 2005.

[381] V. Milirud, L. Fleshel, W. Zhang, G. Jullien, and O. Yadid-Pecht. A WIDE DYNAMIC RANGE CMOS ACTIVE PIXEL SENSOR WITH FRAME DIFFERENCE. In *Int'l Symp. Circuits & Systems (ISCAS)*, pages 588-591, Kobe, Japan, May 2005.

[382] U. Mallik, M. Clapp, E. Choi, G. Cauwenbergs, and R. Etienne-Cummings. Temporal Change ThresholdDetectin Imager. In *Dig. Tech. Papers Int'l Solid-State Circuits Conf. (ISSCC)*, 2005.

[383] V. Brajovic and T. Kanade. Computational Sensor for Visual Tracking with Attention. *IEEE J. Solid-State Circuits*, 33(8): 1199-1207, August 1998.

[384] J. Akita, A. Watanabe, O. Tooyama, M. Miyama, and M. Yoshimoto. An Image Sensor with Fast

Objects' Positions Extraction Function. *IEEE Trans. Electron Devices*, 50(1): 184-190, January 2003.

[385] R. Etienne-Cummings, J. Van der Spiegel, P. Mueller, and M.-Z. Zhang. A foveated silicon retina for two-dimensional tracking. *IEEE Trans. Circuits & Systems II*, 47(6): 504-517, June 2000.

[386] R. D. Burns, J. Shah, C. Hong, S. Pepić, J. S. Lee, R. I. Hornsey, and P. Thomas. Object Location and Centroiding Techniques with CMOS Active Pixel Sensors. *IEEE Trans. Electron Devices*, 50(12): 2369-2377, December 2003.

[387] Y. Sugiyama, M. Takumi, H. Toyoda, N. Mukozaka, A. Ihori, T. Kurashina, Y. Nakamura, T. Tonbe, and S. Mizuno. A High-Speed CMOS Image Sensor with Profile Data Acquiring Function. *IEEE J. Solid-State Circuits*, 40(12): 2816-2823, December 2005.

[388] T. G. Constandinou and C. Toumazou. A Micropower Centroiding Vision Processor. *IEEE J. Solid-State Circuits*, 41(6): 1430-1443, June 2006.

[389] H. Oku, Teodorus, K. Hashimotoa, and M. Ishikawa. High-speed Focusing of Cells Using Depth-From-Diffraction Method. In *Proc. IEEE Int'l Conf. Robotics & Automation (ICRA)*, pages 3636-3641, Orlando, FL, May 2006.

[390] M. F. Land and D.-E. Nilsson. *Animal Eyes*. Oxford University Press, Oxford, UK, 2002.

[391] E. Hecht. *Optics*. Addison-Wesley Pub. Co., Reading, MA, 2nd edition, 1987.

[392] B. A. Wandell. *Foundations of Vision*. Sinauer Associates, Inc., Sunderland, MA, 1995.

[393] R. Wodnicki, G. W. Roberts, and M. D. Levine. A Log-Polar Image Sensor Fabricated in a Standard 1.2-μm ASIC CMOS Process. *IEEE J. Solid-State Circuits*, 32(8): 1274-1277, August 1997

[394] F. Pardo, B. Dierickx, and D. Scheffer. CMOS foveated image sensor: signal scaling and small geometry effects. IEEE Trans. Electron Devices, 44(10): 1731-1737, October 1997.

[395] F. Pardo, B. Dierickx, and D. Scheffer. Space-variant nonorthogonal structure CMOS image sensor design. *IEEE J. Solid-State Circuits*, 33(6): 842-849, June 1998.

[396] F. Saffih and R. Hornsey. Pyramidal Architecture for CMOS Image Sensor. In *IEEE Workshop on Charge-Coupled Devices & Advanced Image Sensors*, Elmau, Germany, May 2003.

[397] K. Yamazawa, Y. Yagi, and M. Yachida. Ominidirectional Imaging with Hyperboloidal Projection. In *Proc. IEEE/RSJ Int'l Conf. Intelligent Robots & Systems*, pages 1029-1034, Yokohama, Japan, July 1993.

[398] J. Ohta, H. Wakasa, K. Kagawa, M. Nunoshita, M. Suga, M. Doi, M. Oshiro, K. Minato, and K. Chihara. A CMOS image sensor for Hyper Omni Vision. *Trans. Inst. Electrical Eng. Jpn.*, E, 123-E (11): 470-476, November 2003.

[399] J. Tanidaa, T. Kumagai, K. Yamada, S. Miyatakea, K. Ishida, T. Morimoto, N. Kondou, D. Miyazaki, and Yoshiki Ichioka. Thin observation module by bound optics (TOMBO): concept and experimental verification. *Appl. Opt.*, 40(11): 1806-1813, April 2001.

[400] S. Ogata, J. Ishida, and H. Koshi. Optical sensor array in an artificial compound eye. *Opt. Eng.*, 33: 3649-3655, November 1994.

[401] J. S. Sanders and C. E. Halford. Design and analysis of apposition compound eye optical sensros. *Opt. Eng.*, 34(1): 222-235, January 1995.

[402] K. Hamanaka and H. Koshi. An artificial compound eye using a microlens array and its application to scale-invariant processing. *Opt. Rev.*, 3(4): 265-268, 1996.

[403] J. Duparré, P. Dannberg, P. Schreiber, A. Bräuer, and A. Tünnermann. Microoptically fabricated artificial apposition compound eye. In *Proc. SPIE*, volume 5301, pages 25-33, San Jose, CA, January 2004.

[404] R. Hornsey, P. Thomas, W. Wong, S. Pepic, K. Yip, and R. Kishnasamy. Electronic compound-eye

image sensor: construction and calibration. In *Proc. SPIE*, volume 5301, pages 13-24, San Jose, CA, January 2004.

[405] J. Tanida, R. Shogenji, Y. Kitamura, K. Yamada, M. Miyamoto, and S. Miyatake. Imaging with an integrated compound imaging system. *Opt. Express*, 11(18): 2109-2117, September 2003.

[406] R. Shogenji, Y. Kitamura, K. Yamada, S. Miyatake, and J. Tanida. Multispectral imaging using compact compound optics. *Opt. Express*, 12(8): 1643-1655, April 2004.

[407] S. Miyatake, R. Shogenji, M. Miyamoto, K. Nitta, and J. Tanida. Thin observation module by bound optics (TOMOBO) with color filters. In *Proc. SPIE*, volume 5301, pages 7-12, San Jose, CA, January 2004.

[408] A. G. Andreou and Z. K. Kalayjian. Polarization Imaging: Principles and Integrated Polarimeters. *IEEE Sensors Journal*, 2(6): 566-576, 2002.

[409] M. Nakagawa. Ubiquitous Visible Light Communications. *IEICE Trans. On Communications*, J88-B(2): 351-359, February 2005.

[410] PhaseSpace, Inc. Phase Space motion digitizer. www.phasespace.com/.

[411] PhoeniX Technologies. The Visualeyes System. ptiphoenix.com/.

[412] D. J. Moore, R. Want, B. L. Harrison, A. Gujar, and K. Fishkin. Implementing Phicons: Combining Computer Vision with InfraRed Technology for Interactive Physical Icons. In *Proc. ACM Symposium on User Interface Software and Technology (UIST)*, pages 67-68, 1999.

[413] J. Rekimoto and K. Nagao. The World through the Computer: Augmented Interaction with RealWorld Environment. In *Proc. ACM Symposium on User Interface Software and Technology (UIST)*, pages 29-36, 1995.

[414] N. Matsushita, D. Hihara, T. Ushiro, S. Yoshimura, and J. Rekimoto. ID Cam: A smart camera for scene capturing and ID recognition. *J. Informatin Process. Soc. Jpn.*, 43(12): 3664-3674, December 2002.

[415] N. Matsushita, D. Hihara, T. Ushiro, S. Yoshimura, J. Rekimoto, and Y. Yamamoto. ID CAM: A Smart Camera for Scene Capturing and ID Recognition. *Proc. IEEE & ACM Int'l Sympo. Mixed & Augmented Reality*, page 227, 2003.

[416] H. Itoh, K. Kosugi, X. Lin, Y. Nakamura, T. Nishimura, K. Takiizawa, and H. Nakashima. Spatial Optical point-to-point communication system for indoor locarion-based information services. In *Proc. ICO Int'l Conf. Optics & Photonics in Technology Frontier*, July 2004.

[417] X. Lin and H. Itoh. *Wireless Personal Information Terminal for Indoor Spatial Optical* Communication System Using a Modified DataSlim2. *Opt. Rev.*, 10(3): 155-160, May-Jun 2003.

[418] K. Kagawa, Y. Maeda, K. Yamamoto, Y. Masaki, J. Ohta, and M. Nunoshita. Optical navigation: a ubiquitous visual remote-control station for home information appliances. In *Proc. Optics Japan*, pages 112-113, 2004. In Japanese.

[419] K. Kagawa, K. Yamamoto, Y. Maeda, Y. Miyake, H. Tanabe, Y. Masaki, M. Nunoshita, and J. Ohta. "Opto-Navi," a Multi-Purpose Visual Remote Controller of Home Information Appliances Using a Custom CMOS Image Sensor. *Forum on Info. Tech. Lett.*, 4: 229-232, April 2005. In Japanese.

[420] K. VKagawa, R. Danno, K. Yamamoto, Y. Maeda, Y. Miyake, H. Tanabe, Y. Masaki, M. Nunoshita, and J. Ohta. Demonstration of mobile visual remote controller "OptNavi" system using home network. *J. Inst. Image Information & Television Eng.*, 60(6): 897-908, June 2006.

[421] www.dlna.org/.

[422] www.echonet.gr.jp/.

[423] www.upnp.org/.

[424] www.havi.org/.

[425] www.irda.org/.

[426] www.bluetooth.com/.

[427] Y. Oike, M. Ikeda, and K. Asada. A smart image sensor with high-speed feeble ID-beacon detection for augmented reality system. In *Proc. European Solid-State Circuits Conf.*, pages 125-128, September 2003.

[428] Y. Oike, M. Ikeda, and K. Asada. Smart Image Sensor with High-speed Highsensitivity ID Beacon Detection for Augmented Reality System. *J. Inst. Image Information & Television Eng.*, 58(6): 835-841, June 2004. In Japanese.

[429] K. Yamamoto, K. Kagawa, Y. Maeda, Y. Miyake, H. Tanabe, Y. Masaki, M. Nunoshita, and J. Ohta. An Opt-Navi system using a custom CMOS image sensor with a function of reading multiple region-of-interests. *J. Inst. Image Information & Television Eng.*, 59(12): 1830-1840, December 2005. In Japanese.

[430] K. Yamamoto, Y. Maeda, Y. Masaki, K. Kagawa, M. Nunoshita, and J. Ohta. A CMOS image sensor with high-speed readout of mulitple region-ofinterests for an Opto-Navigation system. In *Proc. SPIE*, volume 5667, pages 90-97, San Jose, CA, January 2005.

[431] K. Yamamoto, Y. Maeda, Y. Masaki, K. Kagawa, M. Nunoshita, and J. Ohta. A CMOS image sensor for ID detection with high-speed readout of multiple region-of-interests. In *IEEE Workshop on Charge-Coupled Devices & Advanced Image Sensors*, pages 165-168, Karuizawa, Japan, June 2005.

[432] J. R. Barry. *Wireless infrared communications*. Kluwer Academic Publishers, New York, NY, 1994.

[433] www.victor.co.jp/pro/lan/index.html.

[434] J. M. Kahn, R. You, P. Djahani, A. G. Weisbin, B. K. Teik, and A. Tang. Imaging diversity receivers for high-speed infrared wireless communication. IEEE Commun. Mag., 36(12): 88-94, December 1998.

[435] D. C. O'Brien, G. E. Faulkner, E. B. Zyambo, K. Jim, D. J. Edwards, P. Stavrinou, G. Parry, J. Bellon, M. J. Sibley, V. A. Lalithambika, V. M. Joyner, R. J. Samsudin, D. M. Holburn, and R. J. Mears. Integrated Transceivers for Optical Wireless Communications. *IEEE Selcted Topic Quantum Electron.*, 11(1): 173-183, Jan-Feb. 2005.

[436] K. Kagawa, T. Nishimura, T. Hirai, J. Ohta, M. Nunoshita, M. Yamada, Y. Yamasaki, S. Sughishita, and K. Watanabe. A vision chip with a focused highspeed read-out mode for optical wireless LAN. In *Int'l Topical Meeting on Optics in Computing*, pages 183-185, Engelberg, Switzerland, April 2002.

[437] T. Nishimura, K. Kagawa, T. Hirai, J. Ohta, M. Nunoshita, M. Yamada, Y. Yamasaki, S. Sughishita, and K. Watanabe. Design and Fabrication for High-speed BiCMOS Image Sensor with Focused Read-out Mode for Optical Wireless LAN. In *Tech. Digest 7th Optoelectronics & Communications Conf.* (*OECC*2002), pages 212-213, July 2002.

[438] K. Kagawa, T. Nishimura, J. Ohta, M. Nunoshita, Y. Yamasaki, M. Yamada, S. Sughishita, and K. Watanabe. An optical wirelesds LAN system based on a MEMS beam steerer and a vision chip. In *Int'l Conf. Optics-photonics Design & Fabrication* (ODF), pages 135-136, November 2002.

[439] K. Kagawa, T. Nishimura, H. Asazu, T. Kawakami, J. Ohta, M. Nunoshita, Y. Yamasaki, and K. Watanabe. A CMOS Image Sensor Working As High-Speed Photo Receiver as Well as a Position Sensor for Indoor Optical Wireless LAN Systems. In *Proc. SPIE*, volume 5017, pages 86-93, Santa Clara, CA, January 2003.

[440] K. Kagawa, T. Kawakami, H. Asazu, T. Nishimura, J. Ohta, M. Nunoshita, and K. Watanabe. An image sensor based optical receiver fabricated in a standard $0.35\mu m$ CMOS technology for mobile applications. In *IEEE Workshop on Charge-Coupled Devices & Advanced Image Sensors*, Elmau, Germany, May 2003.

[441] K. Kagawa, T. Nishimura, T. Hirai, Y. Yamasaki, J. Ohta, M. Nunoshita, and K. Watanabe. Proposal and Preliminary Experiments of Indoor optical wireless LAN based on a CMOS image sensor with a high-speed readout function enabling a low-power compact module with large downlink capacity. *IEICE Trans. Commun.*, E86-B(5): 1498-1507, May 2003.

[442] K. Kagawa, T. Kawakami, H. Asazu, T. Ikeuchi, A. Fujiuchi, J. Ohta, M. Nunoshita, and K. Watanabe. An indoor optical wireless LAN system with a CMOS-image-sensor-based photoreceiver fabricated in a 0.35-μm CMOS technology. In *Int'l TopicalMeeting Optics in Computing*, pages 90-91, Engelberg, Switzerland, April 2004.

[443] K. Kagawa, H. Asazu, T. Kawakami, T. Ikeuchi, A. Fujiuchi, J. Ohta, M. Nunoshita, and K. Watanabe. Design and fabrication of a photoreceiver for a spatially optical communication using an image sensor. *J. Inst. Image Information & Television Eng.*, 58(3): 334-343, March 2004. In Japanese.

[444] A. Fujiuchi, T. Ikeuchi, K. Kagawa, J. Ohta, and M. Nunoshita. Free-space wavelength-division-multiplexing optical communications using a multichannel photoreceiver. In *Int'l Conf. Optics & Photonics in Technology Frontier (ICO)*, pages 480-481, Chiba, Japan, July 2004.

[445] K. Kagawa, H. Asazu, T. Ikeuchi, Y. Maeda, J. Ohta, M. Nunoshita, and K. Watanabe. A 4-ch 400-Mbps image-sensor-based photo receiver for indoor optical wireless LANs. In *Tech. Dig. 9th Optoelectronics & Communications Conf. (OECC2004)*, pages 822-823, Yokohama, Japan, June 2004.

[446] K. Kagawa, T. Ikeuchi, J. Ohta, and M. Nunoshita. An Image sensor with a photoreceiver function for indoor optical wireless LANs fabricated in 0.8-μm BiCMOS technology. In *Proc. IEEE Sensors*, page 288, Vienna, Austria, October 2004.

[447] B. S. Leibowitz, B. E. Boser, and K. S. J. Pister. A 256-Element CMOS Imaging Receiver for Free-Space Optical Communication. *IEEE J. Solid-State Circuits*, 40(9): 1948-1956, September 2005.

[448] M. Last, B. S. Leibowitz, B. Cagdaser, A. Jog, L. Zhou, B. E. Boser, and K. S. J. Pister. Toward a wireless optical communication link between two small unmanned aerial vehicles. In *Int'l Symp. Circuits & Systems (ISCAS)*, volume 3, pages 930-933, May 2003.

[449] B. Eversmann, M. Jenkner, F. Hofmann, C. Paulus, R. Brederlow, B. Holzapfl, P. Fromherz, M. Merz, M. Brenner, M. Schreiter, R. Gabl, K. Plehnert, M. Steinhauser, G. Eckstein, D. Schmitt-Landsiedel, and R. Thewes. A 128×128 CMOS biosensor array for extracellular recording of neural activity. *IEEE J. Solid-State Circuits*, 38(12): 2306-2317, December 2003.

[450] U. Lu, B. Hu, Y. Shih, C. Wu, and Y. Yang. The design of a novel complementary metal oxide semiconductor detection system for biochemical luminescence. *Biosensors Bioelectron.*, 19(10): 1185-1191, 2004.

[451] H. Ji, P. A. Abshire, M. Urdaneta, and E. Smela. CMOS contact imager for monitoring cultured cells. In *Int'l Symp. Circuits & Systems (ISCAS)*, pages 3491-3495, Kobe, Japan, May 2005.

[452] K. Sawada, T. Ohshina, T. Hizawa, H. Takao, and M. Ishida. A novel fused sensor for photo-and ion-sensing. *Sensors & Actuators B*, 106: 614-618, 2005.

[453] H. Ji, D. Sander, A. Haas, and P. A. Abshire. A CMOS contact imager for locating individual cells. *In Int'l Symp. Circuits & Systems (ISCAS)*, pages 3357-3360, Kos, Greece, May 2006.

[454] J. C. Jackson, D. Phelan, A. P. Morrison, M. Redfern, and A. Mathewson. Characterization of Geiger Mode Avalanche Photodiodes for Fluorescence Decay Measurements. In *SPIE, Photodetector Materials and Devices VII*, volume 4650, San Jose, CA, January 2002.

[455] D. Phelan, J. C. Jackson, R. M. Redfern, A. P. Morrison, and A. Mathewson. Geiger Mode Avalanche Photodiodes for Microarray Systems. In *SPIE, Biomedical Nanotechnology Architectures and*

Applications, volume 4626, San Jose, CA, January 2002.

[456] S. Bellis, J. C. Jackson, and A. Mathewson. Towards a Disposable in vivo Miniature Implantable Fluorescence Detector. In *SPIE, Optical Fibers and Sensors for Medical Diagnostics and Treatment Applications VI*, volume 6083, 2006.

[457] T. Tokuda, A. Yamamoto, K. Kagawa, M. Nunoshita, and J. Ohta. A CMOS image sensor with optical and potential dual imaging function for on-chip bioscientific applications. *Sensors & Actuators A*, 125(2): 273-280, February 2006.

[458] T. Tokuda, I. Kadowaki,. K Kagawa, M. Nunoshita, and J. Ohta. A new imaging scheme for on-chip DNA spots with optical/potential dual-image CMOS sensor in dry situation. *Jpn. J. Appl. Phys.*, 46(4B): 2806-2810, April 2007.

[459] T. Tokuda, K. Tanaka, M. Matsuo, K. Kagawa, M. Nunoshita, and J. Ohta. Optical and electrochemical dual-image CMOS sensor for on-chip biomolecular sensing applications. *Sensors & Actuators A*, 135(2): 315-322, April 2007.

[460] D. C. Ng, T. Tokuda, A. Yamamoto, M. Matsuo, M. Nunoshita, H. Tamura, Y. Ishikawa, S. Shiosaka, and J. Ohta. A CMOS Image Sensor for On-chip in vitro and in vivo Imaging of the Mouse Hippocampus. *Jpn. J. Appl. Phys.*, 45(4B): 3799-3806, April 2006.

[461] D. C. Ng, H. Tamura, T. Tokuda, A. Yamamoto, M. Matsuo, M. Nunoshita, Y. Ishikawa, S. Shiosaka, and J. Ohta. Real Time In vivo Imaging and Measurement of Serine Protease Activity in the Mouse Hippocampus Using a Dedicated CMOS Imaging Device. *J. Neuroscience Methods*, 156(1-2): 23-30, September 2006.

[462] D. C. Ng, T. Tokuda, A. Yamamoto, M. Matsuo, M. Nunoshita, H. Tamura, Y. Ishikawa, S. Shiosaka, and J. Ohta. On-chip biofluorescence imaging inside a brain tissue phantom using a CMOS image sensor for in vivo brain imaging verification. *Sensors & Actuators B*, 119(1): 262-274, November 2006.

[463] D. C. Ng, T. Nakagawa, T. Tokuda, M. Nunoshita, H. Tamura, Y. Ishikawa, S. Shiosaka, and J. Ohta. Development of a Fully Integrated Complementary Metal-Oxide Semiconductor Image Sensor-based Device for Real-time In vivo Fluorescence Imaging inside the Mouse Hippocampus. *Jpn. J. Appl. Phys.*, 46(4B): 2811-2819, April 2007.

[464] T. Sakata, M. Kamahori, and Y. Miyahara. Immobilization of oligonucleotide probes on Si_3N_4 surface and its application to genetic field effect transistor. *Mater. Sci. Eng.*, C24: 827-832, 2004.

[465] M. Urdaneta, M. Christophersen, E. Smela, S. B. Prakash, N. Nelson, and P. Abshire. Cell Clinics Technology Platform for Cell-Based Sensing. *In IEEE/NLM Life Science Systems & Applications Workshop*, Bethesda, MD, July 2006.

[466] K. Hashimoto, K. Ito, and Y. Ishimori. Microfabricated disposable DNA sensor for detection of hepatitis B virus DNA. *Sensors & Actuators B*, 46: 220-225, 1998.

[467] K. Dill, D. D. Montgomery, A. L. Ghindilis, and K. R. Schwarzkopf. Immunoassays and sequence-specific DNA detection on a microchip using enzyme amplified electrochemical detection. *J. Biochem. Biophys. Methods*, 2004.

[468] H. Miyahara, K. Yamashita, M. Takagi, H. Kondo, and S. Takenaka. Electrochemical array (ECA) as and integrated multi-electrode DNA sensor. *T. IEE Jpn.*, 121-E: 187-191, 2004.

[469] A. Frey, M. Schienle, C. Paulus, Z. Jun, F. Hofmann, P. Schindler-Bauer, B. Holzapfl, M. Atzesberger, G. Beer, M. Frits, T. Haneder, H.-C. Hanke, and R. Thewes. A digital CMOS DNA chip. *In Int'l Symp. Circuits & Systems (ISCAS)*, pages 2915-2918, Kobe, Japan, May 2005.

[470] R. D. Frostig, editor. *In Vivo Optical Imaging of Brain Function*. CRC Press, Boca Raton, FL, 2002.

[471] A. W. Toga and J. C. Mazziota. *Brain Mapping: the Methods*. Academic Press, New York, NY, 2nd edition, 2002.

[472] C. Shimizu, S. Yoshida, M. Shibata, K. Kato, Y. Momota, K. Matsumoto, T. Shiosaka, R. Midorikawa, T. Kamachi, A. Kawabe, and S. Shiosaka. Characterization of recombinant and brain neuropsin, a plasticity-related serine protease. *J Biol Chem*., 273: 11189-11196, 1998.

[473] G. Iddan, G. Meron, A. Glukhovsky, and P. Swain. Wireless capsule endoscopy. *Nature*, 405(6785): 417, 2000.

[474] R. Eliakim. Esophageal capsule endoscopy (ECE): four case reports which demonstrate the advantage of bi-directional viewing of the esophagus. *In Int'l Conf. Capsule Endoscopy*, pages 109-110, Florida, 2004.

[475] S. Itoh and S. Kawahito. Frame Sequential Color Imaging Using CMOS Image Sensors. *In ITE Annual Convention*, pages 21-2, 2003. In Japanese.

[476] S. Itoh, S. Kawahito, and S. Terakawa. A 2.6mW 2fps QVGA CMOS Onechip Wireless Camera with Digital Image Transmission Function for Capsule Endoscopes. *In Int'l Symp. Circuits & Systems (ISCAS)*, pages 3353-3356, Kos, Greece, May 2006.

[477] X. Xie, G. Li, X. Chen, X. Li, and Z. Wang. A Low-Power Digital IC Design Inside the Wireless Endoscopic Capsule. *IEEE J. Solid-State Circuits*, 40(11): 2390-2400, 2006.

[478] S. Kawahito, M. Yoshida, M. Sasaki, K. Umehara, D. Miyazaki, Y. Tadokoro, K. Murata, S. Doushou, and A. Matsuzawa. A CMOS Image Sensor with Analog Two-Dimensional DCT-Based Compression Circuits for onechip Cameras. *IEEE J. Solid-State Circuits*, 32(12): 2030-2041, December 1997.

[479] K. Aizawa, Y. Egi, T. Hamamoto, M. Hatoria, M. Abe, H. Maruyama, and H. Otake. Computational image sensor for on sensor compression. *IEEE Trans. Electron Devices*, 44(10): 1724-1730, October 1997.

[480] Z. Lin, M. W. Hoffman, W. D. Leon-Salas, N. Schemm, and S. Balkr. A CMOS Image Sensor for Focal Plane Decomposition. *In Int'l Symp. Circuits & Systems (ISCAS)*, pages 5322-5325, Kobe, Japan, May 2005.

[481] A. Bandyopadhyay, J. Lee, R. Robucci, and P. Hasler. A 8 μW/frame 104×128 CMOS imager front end for JPEG Compression. *In Int'l Symp. Circuits & Systems (ISCAS)*, pages 5318-5312, Kobe, Japan, May 2005.

[482] T. B. Tang, E. A. Johannesen, L. Wang, A. Astaras, M. Ahmadian, A. F. Murray, J. M Cooper, S. P. Beaumont, B. W. Flynn, and D. R. S. Cumming. Toward a Miniature Wireless Integrated Multisensor Microsystem for Industrial and Biomedical Applications. *IEEE Sensors Journal*, 2: 628-635, 2002.

[483] A. Astaras, M. Ahmadian, N. Aydin, L. Cui, E. Johannessen, T.-B. Tang, L. Wang, T. Arslan, S. P. Beaumont, B. W. Flynn, A. F. Murray, S. W. Reid, P. Yam, J. M. Cooper, and D. R. S. Cumming. A miniature integrated electronics sensor capsule for real-time monitoring of the gastrointestinal tract (IDEAS). *In Int'l Conf. Biomedical Eng. (ICBME): The Bio-Era: New Challenges, New Frontiers*, pages 4-7, Singapore, December 2002.

[484] J. G. Linvill and J. C. Bliss. A direct translation reading aid for the blind. *Proc. IEEE*, 54(1): 40-51, January 1966.

[485] J. S. Brugler, J. D. Meindl, J. D. Plummer, P. J. Salsbury, and W. T. Young. Integrated electronics for a reading aid for the blind. *IEEE J. Solid-State Circuits*, SC-4: 304-312, December 1969.

[486] J. D. Weiland, W. Liu, and M. S. Humayun. Retinal Prosthesis. *Annu. Rev. Biomed. Eng.*, 7: 361-404, 2005.

[487] W. H Dobelle, M. G. Mladejovsky, and J. P. Girvin. Artifical vision for the blind: Electrical stimulation of visual cortex offers hope for a functional prosthesis. *Science*, 183(123): 440-

444,1974.

[488] W. H. Dobelle. Artificial vision for the blind by connecting a television camera to the visual cortex. *ASAIO J. (American Soc. Artificial Internal Organs J.)*, 46: 3-9, 2000.

[489] C. Veraart, M. C. Wanet-Defalque, B. Gerard, A. Vanlierde, and J. Delbeke. Pattern recognition with the Optic Nerve Visual Prosthesis. *Artif. Organs*, 11: 996-1004, 2003.

[490] J. Wyatt and J. F. Rizzo. Ocular implants for the blind. *IEEE Spectrum*, 33, 1996.

[491] R. Eckmiller. Learning retinal implants with epiretinal contacts. *Ophthalmic Res.*, 29: 281-289, 1997.

[492] M. Schwarz, R. Hauschild, B. J. Hosticka, J. Huppertz, T. Kneip, S. Kolnsberg, L. Ewe, and H. K. Trieu. Single-Chip CMOS Image Sensors for a Retina Implant System. *IEEE Trans. Circuits & Systems II*, 46(7): 870-877, July 1999.

[493] M. S. Humayun, J. D. Weiland, G. Y. Fujii, R. Greenberg, R. Williamson, J. Little, B. Mech, V. Cimmarusti, G. V. Boeme, G. Dagnelie, and E. de Juan Jr. Visual perception in a blind subject with a chronic microelectronic retinal prosthesis. *Vision Research*, 43: 2573-2581, 2003.

[494] W. Liu and M. S. Humayun. Retinal Prosthesis. *In Dig. Tech. Papers Int'l Solid-State Circuits Conf. (ISSCC)*, pages 218-219, San Francisco, CA, February 2004.

[495] J. F. Rizzo III, J. Wyatt, J. Loewenstein, S. Kelly, and D. Shire. Methods and Perceptual Thresholds for Short-Term Electrical Stimulation of Human Retina with Microelectrode Arrays. *Invest. Ophthalmology & Visual Sci.*, 44(12): 5355-5361, December 2003.

[496] R. Hornig, T. Laube, P. Walter, M. Velikay-Parel, N. Bornfeld, M. Feucht, H. Akguel, G. Rössler, N. Alteheld, D. L. Notarp, J. Wyatt, and G. Richard. A method and technical equipment for an acute human trial to evaluate retinal implant technology. *J. Neural Eng.*, 2(1): S129-S134, 2005.

[497] A. Y. Chow, M. T. Pardue, V. Y. Chow, G. A. Peyman, C. Liang, J. I. Perlman, and N. S. Peachey. Implantation of silicon chip microphotodiode arrays into the cat subretinal space. *IEEE Trans. Neural Syst. Rehab. Eng.*, 9: 86-95, 2001.

[498] A. Y. Chow, V. Y. Chow, K. Packo, J. Pollack, G. Peyman, and R. Schuchard. The artificial silicon retina microchip for the treatment of vision loss from retinitis pigmentosa. *Arch. Ophthalmol.*, 122(4): 460-469, 2004.

[499] E. Zrenner. Will Retinal Implants Restore Vision? *Science*, 295: 1022-1025, February 2002.

[500] E. Zrenner, D. Besch, K. U. Bartz-Schmidt, F. Gekeler, V. P. Gabel, C. Kuttenkeuler, H. Sachs, H. Sailer, B. Wilhelm, and R. Wilke. Subretinal Chronic Multi-Electrode Arrays Implanted in Blind Patients. *Invest. Ophthalmology & Visual Sci.*, 47: E-Abstract 1538, 2006.

[501] D. Palanker, P. Huie, A. Vankov, R. Aramant, M. Seiler, H. Fishman, M. Marmor, and M. Blumenkranz. Migration of Retinal Cells through a Perforated Membrane: Implications for a High-Resolution Prosthesis. *Invest. Ophthalmology & Visual Sci.*, 45(9): 3266-3270, September 2004.

[502] D. Palanker, A. Vankov, P. Huie, and S. Baccus. Design of a high-resolution optoelectronic retinal prosthesis. *J. Neural Eng.*, 2: S105-S120, 2005.

[503] D. Palanker, P. Huie, A. Vankov, A. Asher, and S. Baccus. Towards High-Resolution Optoelectronic Retinal Prosthesis. *BIOS*, 5688A, 2005.

[504] H. Sakaguchi, T. Fujikado1, X. Fang, H. Kanda, M. Osanai, K. Nakauchi, Y. Ikuno, M. Kamei, T. Yagi, S. Nishimura, M. Ohji, T. Yagi, and Yasuo Tano. Transretinal Electrical Stimulation with a Suprachoroidal Multichannel Electrode in Rabbit Eyes. *Jpn. J. Ophthalmol.*, 48(3): 256-261, 2004.

[505] K. Nakauchi, T. Fujikado, H. Kanda, T. Morimoto, J. S. Choi, Y. Ikuno, H. Sakaguchi, M. Kamei, M. Ohji, T. Yagi, S. Nishimura, H. Sawai, Y. Fukuda, and Y. Tano. Transretinal electrical stimulation by an intrascleral multichannel electrode array in rabbit eyes. *Graefe's Arch. Clin. Exp. Ophthalmol.*,

243: 169-174, 2005.

[506] M. Kamei, T. Fujikado, H. Kanda, T. Morimoto, K. Nakauchi, H. Sakaguchi, Y. Ikuno, M. Ozawa, S. Kusaka, and Y. Tano. Suprachoroidal-Transretinal Stimulation (STS) Artificial Vision System for Patients with Retinitis Pigmentosa. *Invest. Ophthalmology & Visual Sci.*, 47: E-Abstract 1537, 2006.

[507] A. Uehara, K. Kagawa, T. Tokuda, J. Ohta, and M. Nunoshita. A CMOS retinal prosthesis with on-chip electrode impedance measurement. *Electron. Lett.*, 40(10): 582-583, March 2004.

[508] Y.-L. Pan, T. Tokuda, A. Uehara, K. Kagawa, J. Ohta, and M. Nunoshita. A Flexible and Extendible Neural Stimulation Device with Distributed Multichip Architecture for Retinal Prosthesis. *Jpn. J. Appl. Phys.*, 44(4B): 2099-2103, April 2005.

[509] A. Uehara, Y.-L. Pan, K. Kagawa, T. Tokuda, J. Ohta, and M. Nunoshita. Micro-sized photo detecting stimulator array for retinal prosthesis by distributed sensor network approach. *Sensors & Actuators A*, 120(1): 78-87, May 2005.

[510] T. Tokuda, Y.-L. Pan, A. Uehara, K. Kagawa, M. Nunoshita, and J. Ohta. Flexible and extendible neural interface device based on cooperative multichip CMOS LSI architecture. *Sensors & Actuators A*, 122(1): 88-98, July 2005.

[511] T. Tokuda, S. Sugitani, M. Taniyama, A. Uehara, Y. Terasawa, K. Kagawa, M. Nunoshita, Y. Tano, and J. Ohta. Fabrication and validation of a multichip neural stimulator for in vivo experiments toward retinal prosthesis. *Jpn. J. Appl. Phys.*, 46(4B): 2792-2798, April 2007.

[512] K. Motonomai, T. Watanabe, J. Deguchi, T. Fukushima, H. Tomita, E. Sugano, M. Sato, H. Kurino, M. Tamai, and M. Koyanagi. Evaluation of Electrical Stimulus Current Applied to Retina Cells for Retinal Prosthesis. *Jpn. J. Appl. Phys.*, 45(4B): 3784-3788, April 2006.

[513] T. Watanabe, K. Komiya, T. Kobayashi, R. Kobayashi, T. Fukushima, H. Tomita, E. Sugano, M. Sato, H. Kurino, T. Tanaka, M. Tamai, , and M. Koyanagi. Evaluation of Electrical Stimulus Current to Retina Cells for Retinal Prosthesis by Using Platinum-Black (Pt-b) Stimulus Electrode Array. In *Ext. Abst. Int'l Conf. Solid State Devices & Materials* (SSDM), pages 890-891, Yokohama, Japan, 2006.

[514] A. Dollberg, H. G. Graf, B. Höfflinger, W. Nisch, J. D. Schulze Spuentrup, K. Schumacher, and E. Zrenner. A Fully Testable Retinal Implant. In *Proc. Int'l. Conf. Biomedical Eng.*, pages 255-260, Salzburg, June 2003.

[515] E. Zrenner. Subretinal chronic multi-electrode arrays implanted in blind patients. In *Abstract Book Shanghai Int'l Conf. Physiological Biophysics*, page 147, Shanghai, China, 2006.

[516] W. Liu, P. Singh, C. DeMarco, R. Bashirullah, M. Humayun, and J. Weiland. Semiconductor-based Implantable Microsystems. In W. Finn and P. LoPresti, editors, *Handbook of Neuroprosthetic Methods*, chapter 6, pages 127-161. CRC Pub. Company, Boca Raton, FL, 2002.

[517] G. F. Poggio, F. Gonzalez, and F. Krause. Stereoscopicmechanisms in monkey visual cortex: binocular correlation and disparity selectivity. *J. Neurosci.*, 8(12): 4531-4550, December 1988.

[518] D. A. Atchison and G. Smith. *Opitcs of the Human Eye*. Butterworth-Heinemann, Oxford, UK, 2000.

[519] B. Sakmann and O. D. Creutzfeldt. Scotopic and mesopic light adaptation in the cat's retina. *Pflögers Arch.*, 313(2): 168-185, June 1969.

[520] J. M. Valeton and D. van Norren. Light adaptation of primate cones: An analysis based on extracellular data. *Vision Research*, 23(12): 1539-1547, December 1983.

[521] V. C. Smith and J. Pokorny. Spectral sensitivity of the foveal cone photopigments between 400 and 500 nm. *Vision Research*, 15(2): 161-171, February 1975.

[522] A. Roorda and D. R. Williams. The arrangement of the three cones classes in the living human eye. *Nature*, 397: 520-522, February 1999.

[523] R. M. Swanson and J. D. Meindl. Ion-implanted complementaryMOS transistors in low-voltage circuits. *IEEE J. Solid-State Circuits*, SC-7(2): 146-153, April 1972.

[524] Y. Taur and T. Ning. *Fundamental of Modern VLSI Devices*. Cambridge University Press, Cambridge, UK, 1988.

[525] N. Egami. Optical image size. *J. Inst. Image Information & Television Eng.*, 56(10): 1575-1576, November 2002.